Eutectic Solvents

Eutectic Solvents

Editors

Piotr Cysewski
Tomasz Jeliński

MDPI • Basel • Beijing • Wuhan • Barcelona • Belgrade • Manchester • Tokyo • Cluj • Tianjin

Editors
Piotr Cysewski
Nicolaus Copernicus University
Poland

Tomasz Jeliński
Nicolaus Copernicus University
Poland

Editorial Office
MDPI
St. Alban-Anlage 66
4052 Basel, Switzerland

This is a reprint of articles from the Special Issue published online in the open access journal *Crystals* (ISSN 2073-4352) (available at: https://www.mdpi.com/journal/crystals/special_issues/Eutectic_Solvents).

For citation purposes, cite each article independently as indicated on the article page online and as indicated below:

LastName, A.A.; LastName, B.B.; LastName, C.C. Article Title. *Journal Name* **Year**, *Volume Number*, Page Range.

ISBN 978-3-0365-0022-5 (Hbk)
ISBN 978-3-0365-0023-2 (PDF)

© 2020 by the authors. Articles in this book are Open Access and distributed under the Creative Commons Attribution (CC BY) license, which allows users to download, copy and build upon published articles, as long as the author and publisher are properly credited, which ensures maximum dissemination and a wider impact of our publications.

The book as a whole is distributed by MDPI under the terms and conditions of the Creative Commons license CC BY-NC-ND.

Contents

About the Editors vii

Preface to "Eutectic Solvents" ix

Piotr Cysewski
Special Issue Editorial: Eutectic Solvents
Reprinted from: *Crystals* 2020, 10, 932, doi:10.3390/cryst10100932 1

Michal Jablonský, Veronika Majová, Jozef Šima, Katarína Hroboňová and Anna Lomenová
Involvement of Deep Eutectic Solvents in Extraction by Molecularly Imprinted Polymers—A Minireview
Reprinted from: *Crystals* 2020, 10, 217, doi:10.3390/cryst10030217 5

Michal Jablonsky, Veronika Majova, Petra Strizincova, Jozef Sima and Jozef Jablonsky
Investigation of Total Phenolic Content and Antioxidant Activities of Spruce Bark Extracts Isolated by Deep Eutectic Solvents
Reprinted from: *Crystals* 2020, 10, 402, doi:10.3390/cryst10050402 17

Michal Jablonský and Jozef Šima
Phytomass Valorization by Deep Eutectic Solvents—Achievements, Perspectives, and Limitations
Reprinted from: *Crystals* 2020, 10, 800, doi:10.3390/cryst10090800 39

Jing Xue, Jing Wang, Daoshuo Feng, Haofei Huang and Ming Wang
Processing of Functional Composite Resins Using Deep Eutectic Solvent
Reprinted from: *Crystals* 2020, 10, 864, doi:10.3390/cryst10100864 75

Yanrong Liu, Zhengxing Dai, Fei Dai and Xiaoyan Ji
Ionic Liquids/Deep Eutectic Solvents-Based Hybrid Solvents for CO_2 Capture
Reprinted from: *Crystals* 2020, 10, 978, doi:10.3390/cryst10110978 95

About the Editors

Piotr Cysewski was born in Poland in 1960. He received his MD and Ph.D. in Chemistry from Nicolaus Copernicus University, Torun, Poland in 1984 and 1988, respectively. In 2013, he was awarded the title of Professor of Chemistry. Currently, he is the General Chair of Physical Chemistry Department, Collegium Medicum, Nicolaus Copernicus University, ul. Kurpińskiego 5, 85-950 Bydgoszcz, Poland. He has published more than 200 scientific papers in the field of computational chemistry applied to biochemically interesting systems. His current research covers such topics as cocrystallization of active pharmaceutical ingredients, structural and energetic diversities of modified double-stranded DNA, including epigenetic modifications and modelling of the structure and properties of deep eutectics.

Education and career

Diploma of Master thesis in Chemistry at the Nicolaus Copernicus University (1984)

Ph.D. at the Nicolaus Copernicus University (1988)

DSci (habilitation) at the Polish Academy of Science, Bioorganic Chemistry Institute in Poznań (2000)

Professor of Chemistry since 2013

Chair of Physical Chemistry Department, Pharmacy Faculty, Collegium Medicum, Nicolaus Copernicus University since 2006

Postdoctoral position at L.P.B.C, C.N.R.S., Thiais (Paris), France (1990)

Invitation of C.N.R.S. L.P.B.C, C.N.R.S., Thiais (Paris), France (1991)

Commission of European Communities exchange 1993 L.P.B.C, C.N.R.S., Thiais (Paris), France (1993)

Invited Professor at Johannes Guttenberg University of Mainz (1994)

Tomasz Jeliński was born in Bydgoszcz, Poland in 1985. He obtained his MD and Ph.D. in Chemistry from University of Science and Technology in Bydgoszcz in 2010 and 2015, respectively. Currently, he is Assistant Professor at the Physical Chemistry Department, Collegium Medicum, Nicolaus Copernicus University. His fields of interest include ecological aspects of epoxy resins, solubility of active pharmaceutical ingredients in various solvent systems, ionic liquids, and natural deep eutectics. In his studies, he focuses on modelling of both the structure and properties of the studied systems as well as their practical applications.

Education and career

Master's degree in Chemical Engineering at the University of Science and Technology (2010)

Research and teaching assistant at Physical Chemistry Department, Pharmacy Faculty, Collegium Medicum, Nicolaus Copernicus University (2011)

Ph.D. in Chemistry at the University of Science and Technology (2015)

Assistant Professor at Physical Chemistry Department, Pharmacy Faculty, Collegium Medicum, Nicolaus Copernicus University (2017)

Preface to "Eutectic Solvents"

Among the plethora of ionic liquid applications, the pharmaceutical domain is an important beneficiary of practical innovative solutions. Particularly important is the development of new forms of drugs in liquid states for direct or encapsulated delivery.

Piotr Cysewski, Tomasz Jeliński
Editors

Editorial

Special Issue Editorial: Eutectic Solvents

Piotr Cysewski

Department of Physical Chemistry, Collegium Medicum, Nicolaus Copernicus University, Kurpińskiego 5, 85-950 Bydgoszcz, Poland; piotr.cysewski@cm.umk.pl

Received: 5 October 2020; Accepted: 9 October 2020; Published: 13 October 2020

Keywords: eutectic solvents; deep eutectic solvents; natural deep eutectic solvents; nanostructured ionic solvents; neoteric solvents; ionic liquids

---

Introduction

Ionic liquids (ILs) is an umbrella term covering a variety of sub-definitions that focus on more specific subjects. This general research area encompasses more than a hundred thousand of papers published since the first discovery of a low melting-temperature organic salt formed by ethanolammonium nitrate in 1888 by German scientists Gabriel and Weiner [1]. Some [2,3] have attributed the starting point to the more recent finding by Paul Walden [4], who discovered the low room temperature melting point of [EtNH$_3$][NO$_3$]. Regardless of the information's actual origin, the potential benefits of lowering the melting points of molten salts was very soon recognized, appreciated, and followed world-wide. Therefore, papers focusing on ionic liquids projects have been repeated like a mantra, thus leading to an explosion of interest in the subject. This explosion becomes immediately understandable when inspecting of the number of papers published every year in the field. Figure 1a documents this eruption of interest in ionic liquids in the 21st century. Additionally, it is worth mentioning the linear growth of the accumulated knowledge over the last decade, as illustrated in Figure 1b. The current year seems to break the trend, but this is probably mainly due to the still-ongoing processing of many papers and omnipresent pandemic restrictions.

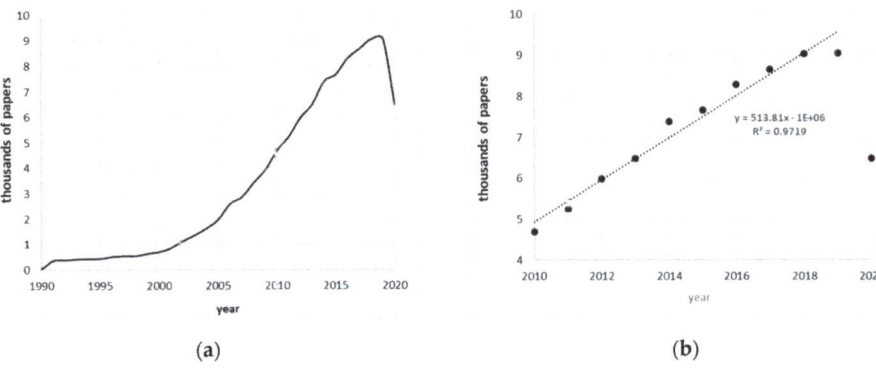

Figure 1. Time evolution of published papers with keyword "ionic liquid" according to Web of Science (updated September 2020) expressed in thousands of papers yearly during (**a**) the last three decades and (**b**) the last decade.

The general explanation for why about 55,900 authors have published more than two papers in the field—written in 23 languages (fortunately with English predomination), coming from 621 countries, founded by over 61 thousand grants, and covering 183 Web of Science Categories—is, of course,

the unique properties of ILs. Compared to others in the large liquidus range, these ILs have ahigh ionic conductivity, high thermal stability, extremely low vapor pressure, high electrical conductivity, large electrochemical window, and the ability to solvate compounds of widely varying polarity [5]. These properties have led to a variety practical applications in many different industrial branches and research areas. The accumulated data on many physical properties for over two thousand available ILs are now freely available in open databases, e.g., ILThermo (v2.0) [6,7].

The first term used in the two-word title of this special issue covers both the outline of the potential systems under interest and the scope of their possible application areas. However, neither are strictly and univocally defined. Indeed, eutectic solvents (ESs), eutectic mixtures (EMs), deep eutectic solvents (DESs), and natural deep eutectic solvents (NADESs) share some physicochemical properties and similar structural foundations. First of all, any eutectic system is a homogeneous mixture of substances melting or solidifying at a lower temperature than the one characterizing any of the constituents and their composition in multicomponent systems. On the other hand, the constituents formulate an eutectic mixture due to mutual interactions between Lewis or Brønsted acids and bases, which are organic or organometallic compounds that adopt anionic and/or cationic forms. Additionally, this is, par excellence, the very definition of ionic liquids. On the other hand, the second term in the title reflects applicability domains as promising functional liquid media (FLM); green, sustainable, and nanostructured ionic solvents (NISs); and neoteric solvents (NSs). From this perspective, the attractiveness of these kind of research topics can be confirmed by the collection of published papers, as exemplified by Figure 2.

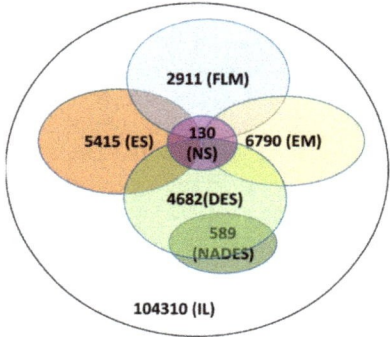

Figure 2. Schematic representation of interests in the specified fields by the overall number of published papers (updated September 2020) focusing on ionic liquids (ILs), functional liquid media (FLM), eutectic solvents (ESs), eutectic mixtures (EMs), and neoteric solvents (NSs).

This special issue represents only small portion of available topics, but it offers two interesting reviews, one mini-review, and one original paper. Laboratory research was significantly prohibited due to global pandemic restrictions. Thus, the collection of new experimental data accumulated despite this disastrous time is of special value. This is why the contribution of Michal Jablonsky is greatly appreciated. I hope that this contribution attracts the attention of a broad range of readers, so I invite open discussion on these new, interesting trends in the field of eutectic mixtures that are practically used as eutectic solvents.

In the first manuscript, which documents the original research of Jablonsky et al. [8], the applicability of deep eutectic solvents for phenolic compound extraction from spruce bark is demonstrated. The authors applied spectrophotometric measurements to quantify the total phenolic content (TPC) and antioxidant activities in extracts of selected eutectic solvents. The extensive set of solvents comprised a variety of molar ratio combinations of choline chloride with lactic acid augmented with 1,3-propanediol, 1,5-pentanediol, 1,4-butanediol, or 1,3-butanediol and water. It was documented that the highest antioxidant activity and radical scavenging activity were found in choline chloride-lactic acid, 1,3-butanediol, and water in a 1:5:1:1 molar ratio. Radical scavenging activity as high as 95% was determined for this eutectic solvent, which was also associated with the highest content of polyphenols in its respective extracts.

Another interesting perspective of eutectic solvent applicability is raised by Xue et al. [9]. In this interesting review, the authors share their perspective on the development and potential benefits of DES-based resin composites. This approach not only adheres to green chemistry policy but can also be used for the further improvement of resin composites. The authors emphasize the necessity of in-depth studies on the intermolecular forces that stabilizing DESs introduce to polymeric matrices. The comprehensive summary of research on the processing of composite resins with DESs is especially valuable from the perspective of low-cost technology for the processing of high-tech products that utilize DES-based composite materials.

The interdisciplinary and far-reaching applications of eutectic solvents are reviewed by Jablonský Šima [10]. Due to the enormous number of experiments on extraction processes, the systematization of the accumulated knowledge of eutectic solvents in the synthetic form is of particular importance. The authors do an excellent job of presenting an organized overview of the use of DESs as extraction agents for the recovery of valuable substances and compounds from original plant biomass. They include waste from its processing and waste from the production and consumption of plant-based food. The alphabetical ordered lists make the data more accessible when information on the extracted particular substances is needed. Furthermore, additional information can be retrieved from provided compilations, including a description of the extracted phytomass, DES composition, extraction conditions, and, of course, all literature sources.

The final article in the issue [11] deals with the subdomain of the extractions represented by molecularly imprinted polymers (MIPs). The authors advocate the application of DESs for the preparation of MIPs by summarizing contemporary achievements in the field. It is of particular importance that the new DESs' applicability is outlined, with focus on potential new breakthrough technology in greener separation, analytical techniques, and the production of MIPs.

In conclusion, it is worth emphasizing the extremely broad and unpredictable span of the potential applications of eutectic solvents in areas that were unimaginable for the pioneers mentioned in the introduction of this editorial. The papers included in this issue are just a few exemplary steps in the ever-expanding possibilities, insights, understandings and inspirations.

Funding: This research received no external funding.

Acknowledgments: As the Guest Editor, I am aware of the unprecedented period since announcing this Special Issue, and I would like to extend my acknowledgement to all the authors for their contributions. Upholding the truth that quality, not quantity, creates real value, I express my recognition of all inspiring papers.

Conflicts of Interest: The author declares no conflict of interest.

References

1. Gabriel, S.; Weiner, J. Ueber einige Abkömmlinge des Propylamins. *Chem. Ber.* **1888**, *21*, 2669–2679. [CrossRef]
2. Welton, T. Ionic liquids: A brief history. *Biophys. Rev.* **2018**, *10*, 691–706. [CrossRef]
3. Smith, E.L.; Abbott, A.P.; Ryder, K.S. Deep Eutectic Solvents (DESs) and Their Applications. *Chem. Rev.* **2014**, *114*, 11060–11082. [CrossRef] [PubMed]
4. Walden, P. Über die Molekulargrösse und elektrische Leitfähigkeit einiger geschmolzener Salze. *Bull. Acad. Imper. Sci. (St. Petersburg)* **1914**, *8*, 405–422.
5. Zhang, S.; Sun, N.; He, X.; Lu, X.; Zhang, X. Physical Properties of Ionic Liquids: Database and Evaluation. *J. Phys. Chem. Ref. Data* **2006**, *35*, 1475–1517. [CrossRef]
6. Kazakov, A.; Magee, J.W.; Chirico, R.D.; Paulechka, E.; Diky, V.; Muzny, C.D.; Kroenlein, K.; Frenkel, M. NIST Standard Reference Database 147: NIST Ionic Liquids Database-(ILThermo); Version 2.0; National Institute of Standards and Technology: Gaithersburg, MD, USA, 2017; p. 20899.

7. Dong, Q.; Muzny, C.D.; Kazakov, A.; Diky, V.; Magee, J.W.; Widegren, J.A.; Chirico, R.D.; Marsh, K.N.; Frenkel, M. ILThermo: A Free-Access Web Database for Thermodynamic Properties of Ionic Liquids. *J. Chem. Eng. Data* **2007**, *52*, 1151–1159. [CrossRef]
8. Jablonsky, M.; Majova, V.; Strizincova, P.; Sima, J.; Jablonsky, J. Investigation of Total Phenolic Content and Antioxidant Activities of Spruce Bark Extracts Isolated by Deep Eutectic Solvents. *Crystals* **2020**, *10*, 402. [CrossRef]
9. Xue, J.; Wang, J.; Feng, D.; Huang, H.; Wang, M. Processing of Functional Composite Resins Using Deep Eutectic Solvent. *Crystals* **2020**, *10*, 864. [CrossRef]
10. Jablonský, M.; Šima, J. Phytomass Valorization by Deep Eutectic Solvents—Achievements, Perspectives, and Limitations. *Crystals* **2020**, *10*, 800. [CrossRef]
11. Jablonský, M.; Majová, V.; Šima, J.; Hroboňová, K.; Lomenová, A. Involvement of Deep Eutectic Solvents in Extraction by Molecularly Imprinted Polymers—A Minireview. *Crystals* **2020**, *10*, 217. [CrossRef]

© 2020 by the author. Licensee MDPI, Basel, Switzerland. This article is an open access article distributed under the terms and conditions of the Creative Commons Attribution (CC BY) license (http://creativecommons.org/licenses/by/4.0/).

Review

Involvement of Deep Eutectic Solvents in Extraction by Molecularly Imprinted Polymers—A Minireview

Michal Jablonský [1,*], Veronika Majová [1], Jozef Šima [2], Katarína Hroboňová [3] and Anna Lomenová [3]

[1] Institute of Natural and Synthetic Polymers, Department of Wood, Pulp and Paper, Faculty of Chemical and Food Technology, Slovak University of Technology in Bratislava, Radlinskeho 9, SK-812 37 Bratislava, Slovakia; veronika.majova@stuba.sk
[2] Department of Inorganic Chemistry, Faculty of Chemical and Food Technology, Slovak University of Technology in Bratislava, Radlinskeho 9, SK-812 37 Bratislava, Slovakia; jozef.sima@stuba.sk
[3] Institute of Analytical Chemistry, Faculty of Chemical and Food Technology, Slovak University of Technology in Bratislava, Radlinskeho 9, SK-812 37 Bratislava, Slovakia; katarina.hrobonova@stuba.sk (K.H.); anna.lomenova@stuba.sk (A.L.)
* Correspondence: michal.jablonsky@stuba.sk

Received: 28 February 2020; Accepted: 18 March 2020; Published: 19 March 2020

Abstract: Substantial research activity has been focused on new modes of extraction and refining processes during the last decades. In this field, coverage of the recovery of bioactive compounds and the role of green solvents such as deep eutectic solvents (DESs) also gradually increases. A specific field of DESs involvement is represented by molecularly imprinted polymers (MIPs). The current state and prospects of implementing DESs in MIPs chemistry are, based on the accumulated experimental data so far, evaluated and discussed in this minireview.

Keywords: deep eutectic solvents; molecularly imprinted polymers; extraction

1. Introduction

Green chemistry and technologies related to it contribute to the improvement of the environment, and also provide a significant economic impact. Remarkable progress has been achieved, mainly in the area of seeking new methods of obtaining chemicals, in particular phytochemicals from plant materials, from natural renewable resources and from waste matter. The goal is to isolate the target compounds or substances selectively, and at the same time eliminate and remove undesirable by-products. Phytochemicals are part of a broad and diverse group of chemical compounds, classified according to their chemical structures and functional properties. As typical representatives, polyphenols, terpenes, amino acids, and proteins can be mentioned [1].

In order to extract desired substances, extraction techniques are currently used—among which the most widely utilized are Soxhlet extraction, accelerated solvent extraction, ultrasound-assisted extraction, microwave-assisted extraction and supercritical fluid extraction [2]. Target chemical compounds differ in their polarity, stability and physical properties, thus rendering a single-step extraction with one solvent for all the compounds from real plant materials generally impossible. To extract, separate and purify the desired substances, several organic solvents are commonly utilized. However, they are often volatile, toxic, flammable, explosive, and their biodegradability is low. That is the rationale behind innovative methods of extraction and separation of analytes in natural materials, which would reduce the consumption of organic solvents, and also improve the efficiency, selectivity and kinetics of extraction. Deep eutectic solvents (DESs) are such alternative solvents.

For isolating, purifying and pre-concentrating individual target substances from primary fractions of extracts, selective sorbents are suitable. Molecularly imprinted polymers (MIPs) have already proven the justification of their use in the isolation of the desired substances. To date, however, the benefits of the combination of DESs and MIPs have not been sufficiently recognized and exploited. The use of DES in MIP synthesis can eliminate some of the disadvantages of traditional procedures (e.g., high volumes of organic solvents) and improve properties of prepared sorbents.

The aim of this minireview is to point out examples of DES usage in MIP synthesis and on the applications of sorbents in extraction procedures for the isolation/purification of substances from complex matrices.

2. Deep Eutectic Solvents

Deep eutectic solvents are mixtures of two or more compounds—hydrogen bond donor (HBD) and hydrogen bond acceptor (HBA)—with a freezing point well below the melting point for any of the original mixture components. From the viewpoint of application, it is preferred that they are liquids at room temperature. The role of HBA is most frequently performed by quaternary ammonium chlorides such as choline chloride (further abbreviated as ChCl) or by amino acids. Urea and imidazole derivatives, amides, alcohols, saccharides or organic carboxylic acids act as HBD. From the chemical point of view, the typical feature of DESs is that HBA and HBD are bonded by the hydrogen bond. When the compounds constituting a DES are exclusively primary metabolites, namely, amino acids, organic acids, sugars, or choline derivatives, the DESs are called natural deep eutectic solvents (NADES). There are also non-eutectic liquid mixtures referred to as low-transition temperature mixtures (LTTMs), composed of high-melting-point starting materials. Since the differences in the properties of the mentioned types of mixtures are from the practical point of view negligible, we will stick to the term DESs. Of the four DESs classes [3], we will mainly deal with the third class of DESs composed of organic constituents. It is worth pointing out that DESs should not be confused with ionic liquids, which are salts in the liquid state with the constituents bonded by the ionic bond.

In comparison with usual solvents, DESs provide many advantages, such as low volatility, low toxicity, miscibility with water, biocompatibility and biodegradability, and low price, and they are also easily prepared with a broad scale of polarity [4–6]. DESs based on ChCl and urea were invented in 2003 [7]. The assessment of their properties (density, viscosity, surface tension, refractive index, pH, etc.) showed their potential to be utilized in industrial applications involving the production of materials with specific properties, the processing of complex materials, and the separation of components from complex mixtures [7].

One of the possibilities of application of DESs lies in obtaining phytochemical extracts for pharmaceutical industry. Another area is the isolation of compounds from biomass, which would be a useful tool for obtaining valuable resources (as raw materials for new products) for various industrial branches, including cosmetic and food industries. There are many combinations of compounds with donor–acceptor properties which may comprise eutectic systems. Besides appropriate physicochemical properties, DESs also offer another benefit—namely its liquid state in a broad interval of temperatures. Before using the prepared mixtures, it is necessary to evaluate the influence of the type and molar ratio of the components on the properties of DESs. Investigating the physical properties of DESs is very important, since they are relatively new systems and have not been examined enough yet. Viscosity and density belong to those properties which vary with temperature and are significant due to diverse applications of DESs [5,8,9]. DESs are often used in a mixture with water, which plays a remarkable role in overcoming the difficulties caused by highly viscous eutectic mixtures. By varying the ratio of HBA and HBD, it is possible to purposefully prepare specific DESs with predefined physicochemical properties, such as melting point, viscosity, conductivity and pH, which are crucial in making the appropriate choices for targeted industrial applications.

3. Extraction by Deep Eutectic Solvents

Significant attention is currently paid to the utilization of DESs for isolating bioactive substances from various resources (biomass, biowaste, food-related waste, plant materials), the extraction of inorganic and organic substances from waste, and materials of biological origin [10]. Results of numerous studies have shown that the usage of "green" solvents often brings about higher extraction efficiency compared to the use of conventional solvents. In recent years, the effort of scientists and technologists has been directed to application of DESs in combination with modern extraction techniques [4,5,11], such as: ultrasound-assisted extraction (UAE) [12–15], negative pressure cavitation extraction (NPC) [13], enzyme assisted extraction (EAE) [3], supercritical fluid extraction (SFE) [16], microwave-assisted extraction (MAE) [9,11], microwave hydrothermal extraction [17], subcritical water extraction [18], and percolation extraction [14]. One of the most important classes of extractable target compounds is polyphenols, exhibiting antioxidant properties, radical scavenging activity, and pharmaceutical and beneficial medical effects [19]. Plant polyphenols comprise the most numerous and widespread group of natural substances isolated from materials of plant origin. Several papers focused on DESs-based extraction of polyphenols, especially flavonoids and phenolic acids from plants, such as Dictamnus albus, Foeniculum vulgare, Origanum majorana, mint, Salvia officinalis [20], Platycladi Cacumen [15], Sophora japonica [21], and others [4,5]. Duan et al. [22] tested five traditional Chinese plants, namely Berberidis Radix, Epimedii Folium, Notoginseng Radix et Rhizoma, Rhei Rhizoma et Radix, and Salviae Miltiorrhizae Radix et Rhizoma in order to evaluate the efficiency of 43 DESs in extraction of alkaloids, flavonoids, saponins, anthraquinones, and phenolic acids. As the results have shown, the extraction efficiency was influenced by all types of DESs. Icariin, a flavonoid, was effectively extracted with proline-containing DESs. Fu et al. [23] investigated the extraction of protocatechuic acid, catechins, epicatechin and caffeic acid from Trachycarpus fortune using DESs as the extraction medium. In order to prepare the DESs, ChCl was mixed with ethylene glycol, glycerol, xylitol, phenol, formic acid, citric acid, oxalic acid and malonic acid. Being environmentally friendly, with low vapor pressure, non-flammability and good thermal stability, DESs proved their high potential for the extraction and purification of polyphenols. The highest extraction yield of protocatechuic acid and epicatechin was achieved using a mixture of ChCl and formic acid in a 1:1 molar ratio at the extraction temperature of 40 °C in a 6-h procedure [23]. Jeong et al. [24] tested 26 DESs, including 9 betaine-based DESs, 8 containing citric acid and 9 containing glycerols, in the process of extraction of catechin from green tea Camellia sinensis. Their results have shown that the mixture containing betaine, glycerol and glucose in molar ratio of 4:20:1 and a pH of 7.16 was the most suitable in comparison with other organic solvents. Škulcová et al. [25] applied various types of DESs to extract compounds from spruce bark (Picea abies). The overall content of polyphenols was determined using the method with Folin–Ciocalteu agent. The polyphenol content in eutectic extracts ranged from 41 to 463 mg of gallic acid equivalent to 100 g of extract. The results of extraction of particular compounds have been thoroughly described in several other publications, as well as in papers by members of the research team [4,26]. The utilization of DESs in polymer synthesis is a new and rapidly developing application area, too. DESs can be used in several phases in the processing, dissolving, extraction, synthesis, and modification of polymers. In recent years, a growing interest concerning DESs' role in the preparation of selective sorbents based on polymers with molecule-imprinted polymers, as well as the utilization of sophisticated approaches towards the molecularly imprinted polymer design, which can significantly reduce the time and cost in optimizing their production.

4. Selective Sorbents Based on Molecularly Imprinted Polymers

Molecularly imprinted polymers (MIPs) are synthetic tailor-made materials with a pre-defined selectivity for a template (frequently target compound), or closely related compounds for which they were designed. These materials are obtained by polymerizing functional and cross-linking monomers around a template molecule, which lead to a highly cross-linked three-dimensional network polymer (Figure 1).

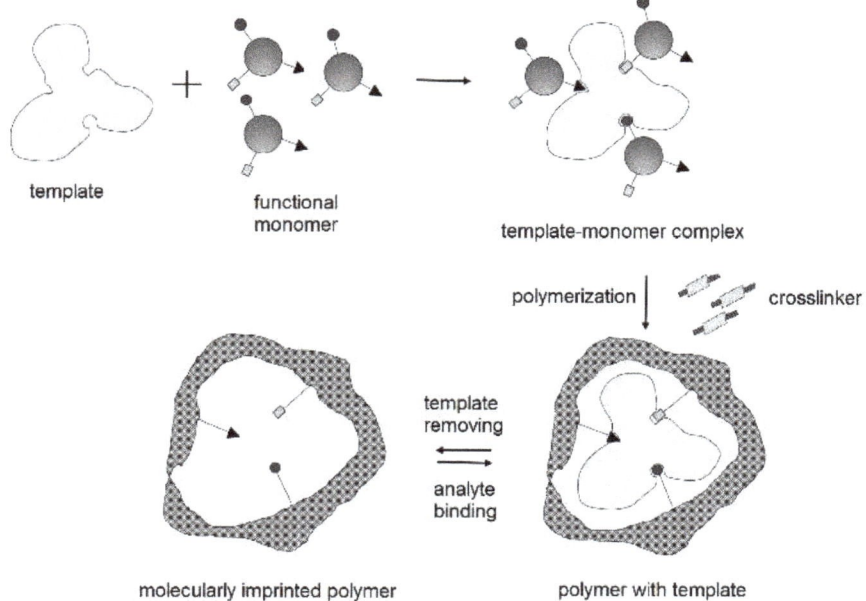

Figure 1. Scheme of MIP preparation [27].

The monomers are selected according to their ability to interact with the functional groups of the template molecule. After polymerization, the template molecules are extracted/removed from the polymeric matrix and binding sites, having their shape, size, and functionalities complementary to the target molecule established. Cavities are able to specifically recognize the target molecule in complex mixtures. The resulting polymers are stable, robust and resistant to organic solvents, high temperatures, and a wide range of pH. In the most common method of preparation, monomers form a complex with a template through covalent or non-covalent interactions. The advantages of the non-covalent approach are the easy formation of the template-monomer complex, the easy removal of the templates from the polymers, fast binding of templates to MIPs and the possibility to prepare for a wide variety of compounds. MIPs are widely applied in the separation, cleaning and pre-concentration of compounds. Conventional MIPs preparation techniques include polymerization in block, precipitation, emulsion, multistep, swelling, suspension and other types of polymerization. The obtained particle size can vary from nano- to micro-particles, from irregular to spherical particles [27,28]. Despite the many advantages of MIPs—such as selectivity, sorption properties, and robustness—they also have disadvantages. When conventional techniques are used, the high quantities of organic solvents as porogens are consumed in preparation process. Water is rarely used, because it can form strong interactions with the template and/or the monomers, and thus destabilize complex formation and also interfere in the formation of specific imprinting sites. Details on the synthesis of MIPs are given elsewhere [29–31]. The use of DESs is an alternative and "green" strategy in MIP preparation, which can eliminate some disadvantages of traditional techniques and solvents. The relationship between DESs and MIPs can be realized in three ways: 1) the usage of a DES in a MIP preparation with the DES acting as a medium/porogen or a reactant incorporated in the MIP; 2) the use of a DES for biomass extraction with subsequent isolation of target compound(s) from the extract by a MIP; 3) the use of a DES as solvent for the extraction of target compound(s) from MIP. While there are a number of examples that meet point 1 (see Table 1), data to meet points 2 and 3 are very rare. In MIPs preparation methodologies, the DESs can be applied as medium or porogen, functional monomer [32–34], MIPs modifier [35,36], or MIPs template [33,37]. Such systems will be abbreviated as DES-MIPs. Some authors postulated that the interaction of a DES with the functional monomer, and/or the surface of a MIP improved the affinity, selectivity and adsorption of target compounds. Such systems will be abbreviated as DES-MIPs. Many publications showed that produced DES-MIPs were suitable for the specific and selective recognition of target compounds in real samples and were characterized by stability, reusability, a high imprinting factor, fast binding kinetics, and high adsorption capacity [38–40]. Some authors also reported the advantages of the DES-MIPs in comparison with MIPs from conventional monomers. The advantages of DESs as monomer compared with conventional monomers are due to their high content of functional groups, allowing unique interactions with template molecules, which result in the higher affinity and selectivity of DES-MIPs. A further advantage is the higher rigidity of DES-MIPs, which can prevent their shrinkage or swelling. Moreover, the liquid phase of DES is advantageous in including the monomer in the bulk of DES or by substituting the media or solvent [36,38,41].

Table 1. The application of deep eutectic solvents (DESs) for molecularly imprinted polymers (MIPs) preparation and extraction/purification procedures.

DES	Molar Ratio	MIPs	Substrate	Target Compounds	Ref.
ChCl:Gl	1:2	Template: chlorogenic acid Monomers: AA Modifier: DES Crosslinker: EGDMA Initiator: AIBN	Honeysuckle	Chlorogenic acid	[32]
ChCl:EG ChCl:Gl ChCl:Bud	1:3 n/n	Template: rutin, scoparone, quercetin Carrier: γ-aminopropyltriethoxysilane-methacrylic acid Monomer: MAA Modifier: DES Crosslinker: EGDMA Initiator: AIBN	Herba Artemisiae Scopariae	Rutin, scoparone, quercetin	[35]
B:EG:W	1:2:1	Template: levofloxacin Monomers: 3-aminopropyltriethoxysilane, MAA, TEOS Modifier: DES Crosslinker: EGDMA Initiator: AIBN Porogen: methanol	Green bean extract	Levofloxacin	[36]
ChCl:EG ChCl:Gl ChCl:Bud ChCl:U ChCl:FA ChCl:AcA ChCl:PA	1:3 n/n	Template: fucodain, alginic acid Carrier: Fe_3O_4@3-aminopropyltriethoxysilane Monomer: MAA Modifier: DES Crosslinker: EGDMA Initiator: AIBN	Seaweed	Fucodain, alginic acid	[37]
ChCl:EG ChCl:Gl ChCl:U ChCl:Bud	1:2 n/n	Template I: tanshinone I, tanshinone IIA, and cryptotanshinone Template II: glycitein, genistein, and daidzein Template III: epicatechin, epigallocatechin gallate, and epicatechin gallate Carrier: Fe_3O_4@SiO_2 Monomers: MAA, DES Crosslinker: EGDMA Initiator: AIBN Porogen: acetonitrile	Salvia miltiorrhiza bunge, Glycine max (Linn.) Merr and green tea	Tanshinone I, tanshinone IIA, and cryptotanshinone from Salvia miltiorrhiza bunge; glycitein, genistein, and daidzein from Glycine max (Linn.) Merr; and epicatechin, epigallocatechin gallate, and epicatechin gallate from green tea	[38]
ChCl:AC	1:2	Template: β-lactoglobulin Carrier: Fe_3O_4@MoS_2 Monomers: DES Crosslinker: EGDMA Initiator: benzoylperoxide, N,N-dimethylaniline Porogen: ethanol:water (9:1)	Milk	β-lactoglobulin, albumin, conalbumin	[39]

Table 1. Cont.

DES	Molar Ratio	MIPs	Substrate	Target Compounds	Ref.
ChCl:MAA	1:2	Template: bovine hemoglobin Carrier: Fe_3O_4@AA Monomers: DES Crosslinker: N,N-methylenebidacrylamide, Initiator: ammonium persulfate, N, N, N´, N´-tetramethylenediamine	Protein solution	Protein	[40]
ChCl:FA ChCl:AcA ChCl:PA ChCl:U	1:2 n/n	Template: laminarin, fucoidan Monomers: MAA, glycidil methacrylate Modifier: DES Crosslinker: EGDMA Initiator: AIBN	Marine kelp	Laminarin, fucoidan	[41]
CfA:ChCl:FA	1:1:0.5 1:2:1 1:3:1.5 1:4:2 1:6:3 1:8:4	Template: levoflocaxin Monomer: DES Crosslinker: EGDMA Initiator: AIBN	Millet extract	Levofloxacin	[42]
B:EG:W	1:2:1	Template: levofloxacin, tetracycline Monomers: 3-aminopropyltriethoxysilane, MAA, TEOS Modifier: DES Crosslinker: EGDMA Initiator: AIBN	Millet extract	Levofloxacin, tetracycline	[43]
ChCl:EG	1:2	Template: gatifloxacin Monomers: 3-aminopropyltriethoxysilane, MAA, TEOS Crosslinker: EGDMA Initiator: AIBN Porogen: DES	Human plasma	Levofloxacin	[44]
ChCl:Gl ChCl:U	(v/v) 0.5:1 1:1 1:2 1:3 1:4 1:5	Template: caffeic acid Monomers: AA, Crosslinker: EGDMA Initiator: AIBN Elution solvent: DES	Hawthorn	Caffeic acid	[45]
ChCl:EG ChCl:Gl ChCl:U	1:2	Template: indomethacin Carrier: mesoporous carbon@MIPS Monomers: MAA, Crosslinker: EGDMA Initiator: AIBN Washing agent: DES	Rat urine	Aristolochic acid I, II	[46]
CfA:ChCl:EG	1:1:1 1:2:2 1:3:3 1:5:5 1:8:8 1:10:10 1:15:15	Template: quercetin Carrier: hexagonal boron nitride Monomers: DES Crosslinker: EGDMA Initiator: AIBN Porogen: methanol	Ginko biloba tea	Quercetin, isorhamnetin, kaempferol	[47]
ChCl:EG ChCl:Gl ChCl:Bud ChCl:U ChCl:FA ChCl:AcA ChCl:PA	1:2	Template: theobromine, theophylline Carrier: Fe_3O_4@MIPs Monomers: MAA Modifier: DES, isopropanol Crosslinker: EGDMA Initiator: AIBN	Green tea	Theobromine, theophylline	[48]
ChCl:OA:EG ChCl:OA:Gl ChCl:OA:PG ChCl:CfA:EG	1:1:1 1:1:2 1:1:3	Template: theophylline, theobromine, (+)-catechin hydrate, caffeic acid Carrier: Fe_3O_4@SiO_2 Monomers: DES Crosslinker: EGDMA Initiator: AIBN Porogen: methanol	Green tea	Theophylline, theobromine, (+)-catechin hydrate, caffeic acid	[49]

Table 1. Cont.

DES	Molar Ratio	MIPs	Substrate	Target Compounds	Ref.
ChCl:EG ChCl:Gl ChCl:PG	1:1	Template: chloramphenicol Monomers: AA Auxiliary monomer: DES Crosslinker: divinilbenzene Initiator: AIBN Porogen: acetonitrile	Milk	Chloramphenicol	[50]
ChCl:Gl	1:2 n/n	Template: chloromycetin, thiamphenicol Monomer: AA Modifier: DES Crosslinker: EGDMA Initiator: 2-methylpropionitrile	Milk	Chloromycetin, thiamphenicol	[51]

The actual researches are focused on new innovative approaches in DES-MIPs preparations DES-MIPs were prepared on the surface of carrier material (magnetite), using ChCl and acrylic acid (1:2) as a functional monomer. This experimental approach avoids the immersion of the template during polymerization and facilitates its removal [33]. Extraction, including solid phase extraction, is a very complex process. For a better understanding of the adsorption of the adsorbate, investigation of the adsorption kinetics is useful. Moreover, the kinetic parameters are useful for designing and modeling the adsorption process, since they can provide information on the number of adsorbed molecules during the adsorption process. To investigate the kinetics of the adsorption process, the pseudo-first order, pseudo-second order models, and intraparticle diffusion model are used [38,39]. The latest results published in the last five years and evaluating the state-of-the-art of methods and technologies applied in the field of DES-MIPs utilization document a wide potential range of these systems in obtaining and/or purifying value-added substances [32,35–51] (Table 1). Extraction methods using DES-MIPs show excellent adsorption ability and selectivity for the selection of templates or target compounds in case studies. In these studies, ChCl acts as HBA in binary or ternary ChCl-based DESs. Research groups of Li and Row [36,43] applied the ternary system containing betaine (HBA), ethylene glycol and water (1:2:1) for extraction of levofloxacin, tetracycline from millet extract [43] and levofloxacin from green bean [36]. Levofloxacin as the target compound was isolated from green been with the recovery reaching 95.2% [36]. From millet, levofloxacin (94.5%), and tetracycline (93.3%), were extracted [43]. Li and Row [42] also studied the application of ternary systems and DES containing caffeic acid:ChCl formic acid in a different molar ratio, and this system as a functional monomer in MIP synthesis. Polymeric sorbent was applied for the purification of levofloxacin. Recovery of levofloxacin for different DES-MIPs (molar ratio for DES: 1:1:0.5; 1:2:1; 1:3:1.5) ranged from 83.2% to 91.3%. Hybrid monomer γ-aminopropyltriethoxysilane-methacrylic acid (KH-550-MAA) was modified by DESs composed of ChCl and ethylene glycol, glycerol or 1,4-butanediol acting as HBD with template (target compound) rutin, scoparone, quercetin were evaluated as more effective from the viewpoint of recoveries of the target compounds compared with hybrid molecular imprinted polymers (HMIPs) modified with ionic liquids [35]. Modified by DES and ionic liquids, the HMIPs were developed for high recognition towards rutin, scoparone, and quercetin in *Herba Artemisiae Scopariae*. The best extraction recoveries were found for the system ChCl:glycerol (1:3)-HMIPs (rutin 92.27%; scoparone 87.51%; quercetin 80.02%). The attention of the authors of papers [50,51] was focused on milk analysis with DES-MIPs. A molar ratio of 1:1 of a mixture of ChCl:ethylene glycol, ChCl:glycerol, or ChCl:propylene glycol [50] or a mixture of ChCl:glycerol (1:2, n/n) [51], with template chloramphenicol, were used in DES-MIPs preparation for milk analysis. These sorbents were applied for extraction of chloromycetin (CHL) and thiamphenicol (THI), which are still used illegally in some animals intended for food production all over the world. DES-MIPs in dispersive liquid–liquid microextraction or in solid-phase extraction show higher recoveries for both analytes/templates (87.23% for CHL; 83.17% for THI /91.23% for CHL; 87.02% for THI, respectively) than MIPs prepared without DES.

The purification of hawthorn extract was achieved by solid-phase extraction process, and SPE recoveries of chlorogenic acid were 72.56%, 64.79%, 69.34% and 60.08% by DES-MIPs, DES-NIPs, MIPs and NIPs, respectively [32]. Non imprinted polymers (NIPs) are synthesized and treated under the same conditions but without the addition of the template. The results showed that strategy of modification of different systems (MIPs) by DES led to improving the properties of polymers due to the controlled morphology and homogeneity of the binding sites. In addition to the possibilities offered by MIPs and DES-MIPs, invention of a new class of MIPs-magnetic MIP (further on MMIPs)-has opened and expanded the possibilities of extraction, isolation and analysis of the desired compounds from materials of biological origin as well as separation of MIPs from reaction systems. DES-MMIPs can contain various magnetic parts as carrier, such as Fe_3O_4@3-aminopropyltriethoxysilane [37], Fe_3O_4@SiO_2 [38]; Fe_3O_4@MoS_2 [39]; Fe_3O_4@acrylic acid [40]; Fe_3O_4 [48]; Fe_3O_4@SiO_2 [49]. MIPs are imprinted on the surface of magnetic parts and have usually a core–shell structure, of which the magnetic phase is the core and the polymeric phase acts as the shell [31]. One of the advantages of MMIPs lies in the fact that after the extraction or elution, particles can be easily separated using an external magnetic field rather than centrifugation or filtration. Fu et al. [39] have renewed and reinforced the interest in the recovery of proteins such as β-lactoglobulin, albumin, conalbumin from milk. The resulting magnetic polymer poly(ChCl-acrylic acid)Fe_3O_4@MoS_2 was in the form of nanospheres. It was characterized by good thermal stability at room temperature, and good adsorption capacity and selectivity for β-lactoglobulin. The strong antibacterial activity of this material was confirmed vs. *S. aureus*, *E. coil* and *B. subtilis*. In the case study [38], MMIPs containing Fe_3O_4@SiO_2 were used. The MIP layer was produced using methacrylic acid (MAA) as a monomer, with the following DESs as porogens:ChCl:urea (DES1); ChCl:ethylene glycol (DES2); ChCl:1,4-butanediol (DES3) and ChCl: glycerol (DES4). Ethylene glycol dimethacrylate (EGDMA) acted as the crosslinker, and 2,2-azobisisobutyronitrile (AIBN) as the initiator. The template role was performed by template I: tanshinone I, tanshinone IIA, or cryptotanshinone; template II: glycitein, genistein, or daidzein; template III: epicatechin, epigallocatechin gallate, or epicatechin gallate. In this study, systems for the extraction of substances from different substrates (*Salvia miltiorrhiza bunge*, *Glycine max (Linn.) Merr* and green tea) by non-DES-MNIP, non-DES-MMIP, DES1-MMIP, DES2-MMIP, DES3-MMIP, DES4-MMIP were compared. The system DES4-MMIP showed the best extraction ability for various substances and substrate. Extraction recoveries reached 85.57% for tanshinone I, 80.58% for tanshinone IIA, 92.12% for cryptotanshinone, 81.65% for glycitein, 87.72% for genistein, 92.24% for daidzein, 86.43% for epicatechin, 80.92% for epigallocatechin gallate, and 93.64% for epicatechin gallate. Furthermore, it was observed that DES-containing polymers showed higher selectivity for nine targets than that of systems without the DESs (for both MMIP and MNIP). It was documented that the MMIPs modified by DES are an innovative approach for the extraction of target substances with a higher selectivity and efficiency in the extraction of target compounds. These results of selective recognition and higher recoveries of polysaccharides were also confirmed in another study using seaweed as a substrate [37]. Taking Fe_3O_4@3-aminopropyltriethoxysilane as a carrier; fucodain and alginic acid as templates; MMA as a monomer; EGDMA as a crosslinker, and AIBN as an initiator, the MMIPs were modified by the DESs—ChCl:ehylene glycol; ChCl:glycerol; ChCl:1,4-butanediol; ChCl:urea; ChCl:formic acid; ChCl acetic acid; and ChCl:propionic acid—were prepared [37]. The selective recognition and separation of proteins by DES-MMIPs was evaluated by Liu et al. [40]. The DES-MMIPs were produced using carrier Fe_3O_4@acrylic acid, ChCl and methacrylic acid as a monomer, N,N-methylenebidacrylamide as crosslinker, ammonium persulfate, N,N,N′,N′-tetramethylenediamine as polymerization initiator, and bovine hemoglobin as a template. The adsorption capacity of the DES-MMIPs and DES-MNIPs with a different amount of monomer were compared, and the results suggested that DES-MIPs had approximately three times higher absorption capacity than DES-MNIPs. The extraction method using DES-MMIPs for the determination of target compounds from green tea was used with carriers Fe_3O_4@MIPs [48] and Fe_3O_4@SiO_2 [49]. In the case of Fe_3O_4@MIPs [48], MIPs modified by binary DESs such as ChCl:ehylene glycol; ChCl:glycerol; ChCl: 1,4-butanediol, ChCl:urea, ChCl:formic acid,

ChCl:acetic acid, ChCl:propionic acid were used. The authors of the paper [49] described magnetic polymers modified by ternary DESs, namely ChCl:oxalic acid:ethylene glycol; ChCl:oxalic acid:glycerol; ChCl:oxalic acid:propylene glycol; ChCl:caffeic acid:ethylene glycol. The resulting modified polymers showed excellent adsorption ability and selectivity. The best system for the extraction of the target substances theophylline, theobromine, (+)-catechin hydrate, and caffeic acid from green tea [49] was Fe_3O_4-ChCl:oxalic acid:polypropylene glycol-MMIPs (5.82; 4.32; 18.36 and 3.96 mg/g, respectively) For the binary system [48], ChCl: urea, and Fe_3O_4@MMIPs [48], the extraction amounts of theobromine and theophylline reached 4.87 mg/g and 5.07 mg/g green tea, respectively.

5. Conclusions

This minireview is focused on the application of deep eutectic solvents (DESs) for the preparation of molecularly imprinted polymers (MIPs). DESs have been designed as an environmentally friendly option for the preparation of MIPs and MMIPs, since these solvents can improve the affinity and selectivity of polymers to a target substance. The actual research showed some innovative approaches in DES-MIPs, or DES-MMIPs preparations and utilization. The applications of DES in the production of MIPs are either as a medium or solvent, as functional monomers, as MIPs modifiers, or as MIPs templates in the processes of extraction, separation or purification technologies. In the production of MIPs, the employment of DESs is based on their use as functional monomers, MIPs modifiers, and MIPs templates, as well as in extraction, separation or purification procedures.

Even though the role of DESs in molecularly imprinted technology is still largely unexplored, a rapid increase in research and the implementation of results can be expected. In particular, the use of these solvents in the production of MIPs can—to a considerable extent—expand their use as a new breakthrough technology in greener separation and analytical techniques.

Author Contributions: M.J., K.H. and J.Š. contributed equally to the conceptualization and design of the work; writing—original draft preparation: M.J., J.Š., V.M., K.H., A.L.; writing—editing: M.J., J.Š., and K.H.; supervision and critical revision of the manuscript: M.J., J.Š.; project administration: M.J.; funding acquisition: M.J. All authors have read and agreed to the published version of the manuscript.

Funding: This work was supported by the Slovak Research and Development Agency under the contracts Nos. APVV-15-0052, APVV-19-0185, APVV-19-0174 and VEGA 1/0403/19, VEGA 1/0412/20. This article was realized also thanks to the support for infrastructure equipment provided by the Operation Program Research and Development for the project "National Center for Research and Application of renewable energy sources" (ITMS 26240120016, ITMS 26240120028), for the project "Competence center for new materials, advanced technologies and energy" (ITMS 26240220073), and for the project "University science park STU Bratislava" (ITMS 26240220084), co-financed by the European Regional Development Fund.

Acknowledgments: The authors would like to acknowledge the financial support of the Slovak Research and Development Agency and support for infrastructure equipment provided by the Operation Program Research and Development.

Conflicts of Interest: The authors declare no conflict of interest. The funders had no role in the design of the study; in the collection, analyses, or interpretation of data; in the writing of the manuscript, or in the decision to publish the results.

Abbreviations

AA	acrylamide
AC	acrylic acid
AcA	acetic acid
AIBN	2,2-azobisisobutyronitrile
B	Betaine
Bud	1,4-butanediol
CfA	caffeic acid
ChCl	choline chloride
CHL	chloromycein
DES	deep eutectic solvent
EG	ethylene glycol
EGDMA	ethylene glycol dimethacrylate

Gl	glycerol
HBA	hydrogen bond acceptor
HBD	hydrogen bond donor
HMIPs	hybrid molecular imprinted polymers
MAA	methacrylic acid
MIP	molecularly imprinted polymers, polymer prepared without addition of template in polymerization mixture
MMIP	magnetic molecularly imprinted polymers
NADES	naturally deep eutectic solvent
NIP	not imprinted polymer
OA	oxalic acid
PA	propionic acid
PG	propylene glycol
TEOS	tetraethoxysilane
THI	thiamphenicol
U	urea
W	water

References

1. Mendonça, P.V.; Lima, M.S.; Guliashvili, T.; Serra, A.C.; Coelho, J.F.J. Deep eutectic solvents (DES): Excellent green solvents for rapid SARA ATRP of biorelevant hydrophilic monomers at ambient temperature. *Polymer* **2017**, *132*, 114–121. [CrossRef]
2. Raks, V.; Al-Suod, H.; Buszewski, B. Isolation, separation, and preconcentration of biologically active compounds from plant matrices by extraction techniques. *Chromatographia* **2018**, *81*, 189–202. [CrossRef] [PubMed]
3. Smith, E.L.; Abbott, A.P.; Ryder, K.S. Deep eutectic solvents (DESs) and their applications. *Chem. Rev.* **2014**, *114*, 11060–11082. [CrossRef] [PubMed]
4. Jablonsky, M.; Skulcova, A.; Malvis, A.; Sima, J. Extraction of value- added components from food industry based and agro-forest biowastes by deep eutectic solvents. *J. Biotechnol.* **2018**, *282*, 46–66. [CrossRef]
5. Jablonský, M.; Šima, J. *Deep Eutectic Solvents in Biomass Valorization*; Spektrum STU: Bratislava, Slovakia, 2019; p. 176.
6. Jablonský, M.; Škulcová, A.; Šima, J. Use of deep eutectic solvents in polymer chemistry—A review. *Molecules* **2019**, *24*, 3978. [CrossRef]
7. Abbott, A.P.; Capper, G.; Davies, D.L.; Rasheed, R.K.; Tambyrajah, V. Novel solvent properties of choline chloride/urea mixtures. *Chem. Commun.* **2003**, *1*, 70–71. [CrossRef]
8. Jablonsky, M.; Haz, A.; Majova, V. Assessing the opportunities for applying deep eutectic solvents for fractionation of beech wood and wheat straw. *Cellulose* **2019**, *26*, 7675–7684. [CrossRef]
9. Jablonsky, M.; Majova, V.; Ondrigova, K.; Sima, J. Preparation and characterization of physicochemical properties and application of novel ternary deep eutectic solvents. *Cellulose* **2019**, *26*, 3031–3045. [CrossRef]
10. Shishov, A.; Bulatov, A.; Locatelli, M.; Carradori, S.; Andruch, V. Application of deep eutectic solvents in analytical chemistry. A review. *Microchem. J.* **2017**, *135*, 33–38. [CrossRef]
11. Ekezie, F.G.C.; Sun, D.W.; Cheng, J.H. Acceleration of microwave-assisted extraction processes of food components by integrating technologies and applying emerging solvents: A review of latest developments. *Trends Food Sci. Technol.* **2017**, *67*, 160–172. [CrossRef]
12. Chen, Z.; Wan, C. Ultrafast fractionation of lignocellulosic biomass by microwave-assisted deep eutectic solvent pretreatment. *Bioresour. Technol.* **2018**, *250*, 532–537. [CrossRef] [PubMed]
13. Bajkacz, S.; Adamek, J. Evaluation of new natural deep eutectic solvents for the extraction of isoflavones from soy products. *Talanta* **2017**, *168*, 329–335. [CrossRef] [PubMed]
14. Bosiljkov, T.; Dujmić, F.; Bubalo, M.C.; Hribar, J.; Vidrih, R.; Brnčić, M.; Zlatic, E.; Redovniković, I.R.; Jokić, S. Natural deep eutectic solvents and ultrasound-assisted extraction: Green approaches for extraction of wine lees anthocyanins. *Food Bioprod. Process.* **2017**, *102*, 195–203. [CrossRef]
15. Zhuang, B.; Dou, L.-L.; Li, P.; Liu, E.-H. Deep eutectic solvents as green media for extraction of flavonoid glycosides and aglycones from *Platycladi Cacumen*. *J. Pharm. Biomed.* **2017**, *134*, 214–219. [CrossRef]
16. Paulaitis, M.; Krukonis, V.J.; Kurnik, R.T.; Reid, R.C. Supercritical fluid extraction. *Rev. Chem. Eng.* **1983**, *1*, 179–250.

17. Yiin, C.L.; Quitain, A.T.; Yusup, S.; Sasaki, M.; Uemura, Y.; Kida, T. Characterization of natural low transition temperature mixtures (LTTMs): Green solvents for biomass delignification. *Bioresour. Technol.* **2016**, *199*, 258–264. [CrossRef]
18. Zhang, Y.; Zhang, Y.; Taha, A.A.; Ying, Y.; Li, X.; Chen, X.; Ma, C. Subcritical water extraction of bioactive components from ginseng roots (Panax ginseng CA Mey). *Ind. Crop. Prod.* **2018**, *117*, 118–127. [CrossRef]
19. Abbas, M.; Saeed, F.; Anjum, F.M.; Afzaal, M.; Tufail, T.; Bashir, M.S.; Ishtiaq, A.; Hussain, S.; Suleria, H.A.R. Natural polyphenols: An overview. *Int. J. Food Prop.* **2017**, *20*, 1689–1699. [CrossRef]
20. Bakirtzi, C.; Triantafyllidou, K.; Makris, D.P. Novel lactic acid-based natural deep eutectic solvents: Efficiency in the ultrasound-assisted extraction of antioxidant polyphenols from common native Greek medicinal plants. *J. Appl. Res. Med. Arom. Plants* **2016**, *3*, 120–127. [CrossRef]
21. Zhao, B.-Y.; Xu, P.; Yang, F.-X.; Wu, H.; Zong, M.-H.; Lou, W.-Y. Biocompatible deep eutectic solvents based on choline chloride: Characterization and application to the extraction of rutin from *Sophora japonica*. *ACS Sustain. Chem. Eng.* **2015**, *3*, 2746–2755. [CrossRef]
22. Duan, L.; Dou, L.-L.; Guo, L.; Li, P.; Liu, E.-H. Comprehensive evaluation of deep eutectic solvents in extraction of bioactive natural products. *ACS Sustain. Chem. Eng.* **2016**, *4*, 2405–2411. [CrossRef]
23. Fu, N.; Lv, R.; Guo Z.; Guo, Y.; You, X.; Tang, B.; Han, D.; Yan, H.; Row, K.H. Environmentally friendly and non-polluting solvent pretreatment of palm samples for polyphenol analysis using choline chloride deep eutectic solvents. *J. Chromatogr. A* **2017**, *1492*, 1–11. [CrossRef] [PubMed]
24. Jeong, K.M.; Ko, J.; Zhao, J.; Jin, Y.; Han, S.Y.; Lee, J. Multi-functioning deep eutectic solvents as extraction and storage media for bioactive natural products that are readily applicable to cosmetic products. *J. Clean. Prod.* **2017**, *151*, 87–95. [CrossRef]
25. Škulcova, A.; Haščičová, Z.; Hrdlička, L.; Šima, J.; Jablonský, M. Green solvents based on choline chloride for the extraction of spruce bark (*Picea abies*). *Cell. Chem. Technol.* **2017**, *52*, 3–4.
26. Jablonský, M.; Škulcová, A.; Kamenská, L.; Vrška, M.; Šima, J. Deep eutectic solvents: Fractionation of wheat straw. *BioResources* **2015**, *10*, 8039–8047. [CrossRef]
27. Lomenova, A.; Hroboňová, K. Polyméry s odtlačkami molekúl ako chirálne stacionárne fázy v HPLC. *Chem. Listy* **2019**, *113*, 156–164.
28. Machyňáková, A.; Hroboňová, K. Možnosti prípravy polymérov s odtlačkami molekúl. *Chem. Listy* **2016**, *110*, 609–615.
29. BelBruno, J.J. Molecularly imprinted polymers. *Chem. Rev.* **2018**, *119*, 94–119. [CrossRef]
30. Cormack, P.A.; Elorza, A.Z. Molecularly imprinted polymers: Synthesis and characterisation. *J. Chromatogr. B* **2004**, *804*, 173–182. [CrossRef]
31. Huang, S.; Xu, J.; Zeng, J.; Zhu, F.; Xie, L.; Ouyang, G. Synthesis and application of magnetic molecularly imprinted polymers in sample preparation. *Anal. Bioanal. Chem.* **2018**, *410*, 3991–4014. [CrossRef]
32. Li, G.; Wang, W.; Wang, Q.; Zhu, T. Deep eutectic solvents modified molecular imprinted polymers for optimized purification of chlorogenic acid from honeysuckle. *J. Chromatogr. Sci.* **2016**, *54*, 271–279. [CrossRef] [PubMed]
33. Fu, N.; Li, L.; Liu, X.; Fu, N.; Zhang, C.; Hu, L.; Li, D.; Tang, B.; Zhu, T. Specific recognition of polyphenols by molecularly imprinted polymers based on a ternary deep eutectic solvent. *J. Chromatogr. A* **2017**, *1530*, 23–34. [CrossRef] [PubMed]
34. Ma, W.; Dai, Y.; Row, K.H. Molecular imprinted polymers based on magnetic chitosan with different deep eutectic solvent monomers for the selective separation of catechins in black tea. *Electrophoresis* **2018**, *39*, 2039–2046. [CrossRef] [PubMed]
35. Li, G.; Ahn, W.S.; Row, K.H. Hybrid molecularly imprinted polymers modified by deep eutectic solvents and ionic liquids with three templates for the rapid simultaneous purification of rutin, scoparone, and quercetin from *Herba Artemisiae Scopariae*. *J. Sep. Sci.* **2016**, *39*, 4465–4473. [CrossRef] [PubMed]
36. Li, X.; Row, K.H. Application of deep eutectic solvents in hybrid molecularly imprinted polymers and mesoporous siliceous material for solid-phase extraction of levofloxacin from green bean extract. *Anal. Sci.* **2017**, *33*, 611–617. [CrossRef]
37. Li, G.; Row, K.H. Magnetic molecularly imprinted polymers for recognition and enrichment of polysaccharides from seaweed. *J. Sep. Sci.* **2017**, *40*, 4765–4772. [CrossRef]

38. Li, G.; Wang, X.; Row, K.H. Magnetic molecularly imprinted polymers based on silica modified by deep eutectic solvents for the rapid simultaneous magnetic-based solid-phase extraction of *Salvia miltiorrhiza bunge*, *Glycine max (Linn.) Merr* and green tea. *Electrophoresis* **2018**, *39*, 1111–1118. [CrossRef]
39. Fu, N.; Li, L.; Liu, K.; Kim, C.K.; Li, J.; Zhu, T.; Li, J.; Tang, B. A choline chloride-acrylic acid deep eutectic solvent polymer based on Fe_3O_4 particles and MoS_2 sheets (poly (ChCl-AA DES)@ Fe_3O_4@ MoS_2) with specific recognition and good antibacterial properties for β-lactoglobulin in milk. *Talanta* **2019**, *197*, 567–577. [CrossRef]
40. Liu, Y.; Wang, Y.; Dai, Q.; Zhou, Y. Magnetic deep eutectic solvents molecularly imprinted polymers for the selective recognition and separation of protein. *Anal. Chim. Acta* **2016**, *936*, 168–178. [CrossRef]
41. Li, G.; Dai, Y.; Wang, X.; Row, K. Molecularly imprinted polymers modified by deep eutectic solvents and ionic liquids with two templates for the simultaneous solid-phase extraction of fucoidan and laminarin from marine kelp. *Anal. Lett.* **2019**, *52*, 511–525. [CrossRef]
42. Li, X.; Row, K.H. Application of novel ternary deep eutectic solvents as a functional monomer in molecularly imprinted polymers for purification of levofloxacin. *J. Chromatogr. B* **2017**, *1068*, 56–63. [CrossRef] [PubMed]
43. Li, X.; Row, K.H. Purification of antibiotics from the millet extract using hybrid molecularly imprinted polymers based on deep eutectic solvents. *RSC Adv.* **2017**, *7*, 16997–17004. [CrossRef]
44. Meng, J.; Wang, X. Microextraction by Packed Molecularly Imprinted Polymer Combined Ultra-High-Performance Liquid Chromatography for the Determination of Levofloxacin in Human Plasma. *J. Chem.* **2019**, *2019*, 1–9. [CrossRef]
45. Li, G.; Tang, W.; Cao, W.; Wang, Q.; Zhu, T. Molecularly imprinted polymers combination with deep eutectic solvents for solid-phase extraction of caffeic acid from hawthorn. *Chin. J. Chromatogr.* **2015**, *33*, 792–798. [CrossRef]
46. Ge, Y.-H.; Shu, H.; Xu, X.-Y.; Guo, P.-Q.; Liu, R.-L.; Luo, Z.-M.; Chang, C.; Fu, Q. Combined magnetic porous molecularly imprinted polymers and deep eutectic solvents for efficient and selective extraction of aristolochic acid I and II from rat urine. *Mater. Sci. Eng. C* **2019**, *97*, 650–657. [CrossRef]
47. Li, X.; Row, K.H. Preparation of deep eutectic solvent-based hexagonal boron nitride-molecularly imprinted polymer nanoparticles for solid phase extraction of flavonoids. *Microchim. Acta* **2019**, *186*, 753. [CrossRef]
48. Li, G.; Wang, X.; Row, K.H. Magnetic solid-phase extraction with Fe3O4/molecularly imprinted polymers modified by deep eutectic solvents and ionic liquids for the rapid purification of alkaloid isomers (theobromine and theophylline) from green tea. *Molecules* **2017**, *22*, 1061. [CrossRef]
49. Li, G.; Row, K.H. Ternary deep eutectic solvent magnetic molecularly imprinted polymers for the dispersive magnetic solid-phase microextraction of green tea. *J. Sep. Sci.* **2018**, *41*, 3424–3431. [CrossRef]
50. Tang, W.; Gao, F.; Duan, Y.; Zhu, T.; Ho Row, K. Exploration of deep eutectic solvent-based molecularly imprinted polymers as solid-phase extraction sorbents for screening chloramphenicol in milk. *J. Chrom. Sci.* **2017**, *55*, 654–661. [CrossRef]
51. Li, G.; Zhu, T.; Row, K.H. Deep eutectic solvents for the purification of chloromycetin and thiamphenicol from milk. *J. Sep. Sci.* **2017**, *40*, 625–634. [CrossRef]

© 2020 by the authors. Licensee MDPI, Basel, Switzerland. This article is an open access article distributed under the terms and conditions of the Creative Commons Attribution (CC BY) license (http://creativecommons.org/licenses/by/4.0/).

Article

Investigation of Total Phenolic Content and Antioxidant Activities of Spruce Bark Extracts Isolated by Deep Eutectic Solvents

Michal Jablonsky [1,*], Veronika Majova [1], Petra Strizincova [1], Jozef Sima [2] and Jozef Jablonsky [3]

[1] Institute of Natural and Synthetic Polymers, Department of Wood, Pulp and Paper, Faculty of Chemical and Food Technology, Slovak University of Technology in Bratislava. Radlinskeho 9, SK-812 37 Bratislava, Slovakia; veronika.majova@stuba.sk (V.M.); petra.strizincova@stuba.sk (P.S.)
[2] Department of Inorganic Chemistry, Faculty of Chemical and Food Technology, Slovak University of Technology in Bratislava, Radlinskeho 9, SK-812 37 Bratislava, Slovakia; jozef.sima@stuba.sk
[3] St. Elisabeth University of Health Care and Social Work Bratislava, Nám.1.mája 1, 810 00 Bratislava, Slovakia; jablonskyjozef@gmail.com
* Correspondence: michal.jablonsky@stuba.sk

Received: 20 March 2020; Accepted: 14 May 2020; Published: 16 May 2020

Abstract: Extracts from spruce bark obtained using different deep eutectic solvents were screened for their total phenolic content (TPC) and antioxidant activities. Water containing choline chloride-based deep eutectic solvents (DESs) with lactic acid and 1,3-propanediol, 1,3-butanediol, 1,4-butanediol, and 1,5-pentanediol, with different molar ratios, were used as extractants. Basic characteristics of the DESs (density, viscosity, conductivity, and refractive index) were determined. All the DESs used behave as Newtonian liquids. The extractions were performed for 2 h at 60 °C under continuous stirring. TPC was determined spectrophotometrically, using the Folin-Ciocalteu reagent, and expressed as gallic acid equivalent (GAE). The antioxidant activity was determined spectrophotometrically by 2,2-diphenyl-1-picrylhydrazyl (DPPH) radical scavenging assay. The TPC varied from 233.6 to 596.2 mg GAE/100 g dry bark; radical scavenging activity (RSA) ranged between 81.4% and 95%. This study demonstrated that deep eutectic solvents are suitable solvents for extracting phenolic compounds from spruce bark.

Keywords: deep eutectic solvents; phenolic compounds; antioxidant activity; spruce bark; extraction

---

1. Introduction

Valorization of biomass, bio-waste, and food-related wastes (hereinafter biomass), including extraction of value-added compounds from these sources, represents a dynamically developing area of research and technology [1–34]. In the field of biomass valorization, a number of green extraction methods have been applied, and their results and usability have been reviewed (for the latest reviews see, [1,2,13,16–20]).

A substantial body of research has focused on the new modes of extraction and refining processes during the last decades. Biomass contains many exceptional and especially health-promoting substances. The most important potential uses of compounds extracted from biomass includes pharmaceutical and biomedical applications, and applications in the food industry as additives and functional substances, as well as nutraceuticals used to enhance food quality or in gastronomy [1–3].

In the last decade, several research teams have published data focused on the purposeful processing of many kinds of biomass. Examples include *Pseudowintera colorata* (horopito), a plant native to New Zealand, herbal tea [21], olive, soy, peanuts, corn, and sunflower oil [22], olive cake, onion seed, and by-products from the tomato and pear canning industries [23], pomelo [24], rice straw [25], wood, straw, pulp [26–28], bark of spruce and other tree species [29–31], corn stover, switchgrass, saffron wastes [32], *Miscanthus* [33], and coffee and cocoa by-products [34]. In the field of biomass pretreatment and delignification with DESs, several DESs have been investigated, and the results of their usability have been examined [13]. One of the main tasks of this industry is the separation of the lignin and the cellulosic fractions of the biomass. There are several methods and procedures to change the composition of the original lignocellulosic matrix aimed at eliminating one of its components (lignin or polysaccharides), thus obtaining new products (pulp, microcrystalline cellulose, nanocellulose etc.), and valorizing the biomass. These include processes such as solubilisation, extraction, fractionation, deconstruction, delignification, and post-delignification [1,13,28,33–41]. Value-added substances and compounds are isolated from biomass by extraction techniques using predominantly water and common organic solvents, and, to a minor extent, eco-friendly green solvents represented by deep eutectic solvents [13]. To reach high extraction yield of target compounds, various extraction techniques have been developed, the most frequently used being supercritical fluid extraction, pressurized liquid extraction, and ultrasound-assisted extraction [7].

One of the most important classes of extractable target compounds is represented by polyphenols, which exhibit, *inter alia*, antioxidant properties, mainly due to their radical scavenging activity [11,12,14,15,42–49]. These substances present exceptional properties, exerting antagonist, antiallergic, antiangiogenetic, antiatherosclerotic, anticancer, antidiarrheal, antihypertensive, anti-inflammatory, antimicrobial, antimutagenic, antimycotical, antineoplastic, antioxidant, antiproliferative, antiseptic, antitumor, antiviral, cytotoxic, estrogenic, fungicidal, hepatoprotective, insecticidal, neuroprotective, and pharmacokinetic activities [3]. In vitro antioxidant activity of bark extract is described by Selvasundhari et al. (2014) [50] and Patrick et al. (2016) [51]. The effectiveness of the extraction is quantified through total extract yield, total phenolic content (TPC), total flavonoid content, and total antioxidant capacity [52]. Particular attention is devoted to *trans*-resveratrol (*trans*-RSV), which is considered a powerful compound capable to improve health and prevent chronic diseases in humans, protecting against some neurodegenerative diseases, obesity and diabetes, high blood pressure, as well as cancer and osteoporosis [7]. In Table 1 the total phenolic content and the antioxidant activity for spruce and pine bark extracts obtained by supercritical fluid extraction, pressurized liquid extraction, ultrasound-assisted extraction, soxhlet extraction, and ohmic heating extraction techniques are reported.

Table 1. Total phenolic content (TPC) and antioxidant activity of spruce and pine extracts.

Extraction	TPC (mg GAE/g Dry Extract)	ABTS (mg TEs/g Dry Extract)	FRAP (µmol FeSO$_4$·7H$_2$O/g Dry Extract)	Ref.
SFE_10 % conc. of ethanol, *v/v*	0.77	2.48	8.31	[7]
SFE_20 % conc. of ethanol, *v/v*	1.24	3.08	10.01	[7]
SFE_40 % conc. of ethanol, *v/v*	2.50	5.29	25.49	[7]
PLE_ethanol	33.45	69.87	389.10	[7]
PLE_ethanol	46.32	257.11	506.10	[7]
UAE_ethanol	54.97	128.47	580.25	[7]
SFE_ ethanol	6.11–11.30	0.68–0.79 *		[53]
Soxhlet extraction_n-hexane		8.3 *		[54]
Soxhlet extraction_n-hexane		4.5 *		[54]
ASE_n-hexane		15 *		[55]
Ohmic heating extraction_water ***		136–156 **		[56]

Table 1. Cont.

Extraction	TPC (mg GAE/g Dry Extract)	ABTS (mg TEs/g Dry Extract)	FRAP (µmol FeSO$_4$·7H$_2$O/g Dry Extract)	Ref.
Ohmic heating extraction_50 % conc. of ethanol, v/v, ***		807–990 **		[56]
Extraction_50 % conc. of ethanol, v/v, ***		394–444 **		[56]
Extraction_water, ***		111–120 **		[56]

ASE: accelerated solvent extraction; PLE: pressurized liquid extraction; SFE: supercritical fluid extraction; UAE: ultrasound-assisted extraction; TEs: Trolox equivalent; FeSO$_4$·7H$_2$O: ferrous sulfate heptahydrate; ABTS: 2,20-azinobis (3-ethylbenzothiazoline-6-sulfonic acid); FRAP: ferric reducing antioxidant power; TPC: total phenolic content, * - mmol TEs/g dry extract, ** - µmol TE/ g dry bark; *** - pine bark.

This paper is devoted to three mutually overlapping research topics: softwood bark as the object of processing; polyphenols as the extracted antioxidants; and deep eutectic solvents as extractants.

The focus on softwood bark is explained by the fact that the annual volume of harvested soft woods in Central and Northern Europe is about 25×10^7 m^3, of which ca. 10% is bark, which is currently disposed of or burned for energy recovery. Thus, Norway spruce (Picea abies [Karst.]) bark can be regarded as a largely available source of condensed polyphenols in Europe [5–10]. Meanwhile, the rationale behind the choice of polyphenols lies in their mentioned properties and usability.

Deep eutectic solvents (DESs) are mixtures of two or more components—a hydrogen bond donor (HBD) and a hydrogen bond acceptor (HBA)—which can bond with each other to form a eutectic mixture having a lower final melting point relative to the melting points of the HBA and HBD [4]. From the practical point of view, it is an advantage if a formed DES is liquid at room temperature.

When the compounds that constitute the DES are exclusively primary metabolites, namely, amino acids, organic acids, sugars, or choline derivatives, the DESs are called natural deep eutectic solvents (NADESs). The term "low-transition temperature mixtures" is used for both types of eutectic mixtures (DESs and NADESs), as well as for liquids composed of natural high-melting-point starting materials, which are not eutectic. Common features of the mentioned solvents and slight differences between them are discussed elsewhere [13], in this work we will stick to the most frequently used term "deep eutectic solvents" (DESs).

Contrary to the majority of organic solvents, which are inflammable liquid substances with relatively high vapor pressure, low viscosity, frequently considerable toxicity for living organisms, and with a negative impact on the environment, DESs have attractive physicochemical properties, such as fire resistance, negligible vapor pressure, miscibility with water, and liquid state in a wide temperature range. Being multi-component systems, DESs offer significant advantages over conventional organic solvents; their structure may be modified by the selection of solvent-forming components, as well as by the molar ratio of the components participating in hydrogen bond formation. That is why their properties (e.g., freezing temperature, viscosity, conductivity, refractive index, density, and pH) are significantly influenced by the molar composition of the compounds in the mixture and can be purposefully modified or even optimized [13].

Various methods and conditions have been investigated to extract polyphenols from softwood bark, and, depending on the method applied, different TPC values have been reached for the same bark sample [43]. The effect of temperature was documented by Lazar et al., [57] who reached TPCs of 37.3 mg and 43.1 mg GAE/g spruce bark at 45 °C and 60 °C, respectively. Conde et al. [58] investigated the effect of temperature and pressure on the extractive yield and the total amount of phenolic compounds from maritime pine and beech wood under conditions of 10–25 MPa, 30–50 °C, supercritical CO$_2$ with 10% ethanol. The highest extraction yield (6.1 g extract/100 g wood) was reached at 30 °C and 15 MPa, and TPC 7.6 g GAE per 100 g extract at 50 °C and 25 MPa.

Jablonsky et al. [3] summarized the properties of 237 compounds from studies on coniferous bark extraction. The authors reviewed the activities of various substances isolated from the bark: cytotoxic (25 identified substances); antioxidant (26 substances); antibacterial (42 substances); anti-inflammatory (10 substances); antimutagenic (5 substances); pharmacokinetic (5 substances); pheromone (10 substances); and inhibitors (22 substances).

Škulcová et al. [29] applied different types of DESs to extract compounds from spruce bark. The extracts from spruce bark showed increased antioxidant activity compared with the corresponding pure DES. The polyphenols content in eutectic extracts ranged from 41 to 463 mg of gallic acid equivalent (GAE) per 100 g of extract. The highest levels of polyphenols were achieved using the following ChCl-based DESs: lactic acid (463 mg GAE/100 g extract); glycolic acid (398 mg GAE/100 g extract); malonic acid (209 mg GAE/100 g extract); tartaric acid (198 mg GAE/100 g extract); oxalic acid (191 mg GAE/100 g extract); citric acid (119 mg GAE/100 g extract); glycerol (82 mg GAE/100 g extract); maleic acid (52 mg GAE/100 g extract); and the lowest level of polyphenols using ChCl with malic acid extract (41 mg GAE/100 g extract). Chupin et al. [59] obtained 18.07 ± 3.82 mg GAE/g bark when extracting pine bark with 80% aqueous ethanol by MAE. The water/ethanol mixture extracted the highest content of phenolic substances (73.48 ± 1.84 mg GAE/g DM), followed by ethanol and then by water (63.38 ± 1.26 mg GAE/g DM and 50.09 ± 4.70 mg GAE/g DM, respectively) [60], where DM is dry matter. Jablonsky et al. [61] summarized the pharmacokinetic properties of biomass-extracted substances isolated by green solvents.

In this study, three-component systems with choline chloride, lactic acid, and water, as well as four-component systems comprising water, choline chloride, lactic acid in different combinations with 1,3-propanediol, 1,3-butanediol, 1,4-butanediol, or 1,5-pentanediol, with different molar ratios, were used to extract phenolic compounds from spruce (*Picea abies*) bark. The DESs used were first prepared and their physico-chemical properties described in Jablonsky et al. (2019) [28]. The novelty of this paper lies in two factors: a new source of antioxidant compounds was used (*Picea abies*); and DESs were used as green solvents to isolate antioxidant compounds. Moreover, antioxidant activity and total phenolic contents of different extracts from spruce bark were determined.

2. Materials and Methods

2.1. Chemicals

All reagents, standards, and solvents were of analytical grade. Choline chloride (ChCl) (≥ 98.0%), 1,3-propanediol (98%), 1,3-butanediol (≥ 99.5%), 1,4-butanediol (≥ 99.0%), 1,5-pentanediol (≥ 96.0%), Folin-Ciocalteu reagent, 2,2-diphenyl-1-picrylhydrazyl radical were purchased from Sigma-Aldrich (Germany). Lactic acid (LacA) 90.0% solution was obtained from VWR International (Bratislava, Slovakia). Choline chloride was dried under vacuum. The other chemicals were used as supplied without further purification.

2.2. Preparation of Deep Eutectic Solvents

The DESs were prepared by mixing and stirring the corresponding components in a water bath (60 °C; 30 min) to form a homogeneous liquid. Key information about the DESs used is summarized in Table 2. The main characteristics of the DESs are gathered in Table 3.

Table 2. Prepared deep eutectic solvents (DESs), molar ratios of their components, and viscosity.

Sample	Component A	Component B	Component C	Component D	Molar Ratio	Water Content (%)	Viscosity at 60 °C (mPa S)
DES1	ChCl	LacA	-	Water	1:2:0.96	5.4	31.1
DES2	ChCl	LacA	-	Water	1:3:0.97	6.4	26.1
DES3	ChCl	LacA	-	Water	1:4:0.99	7.1	21.3
DES4	ChCl	LacA	-	Water	1:5:0.98	7.5	18.9
DES5	ChCl	LacA	1,3-propanediol	Water	1:1:1:0.92	3.4	25.5
DES6	ChCl	LacA	1,3-propanediol	Water	1:2:1:0.95	4.8	18.2
DES7	ChCl	LacA	1,3-propanediol	Water	1:3:1:0.91	5.6	15.9
DES8	ChCl	LacA	1,3-propanediol	Water	1:4:1:0.92	6.4	15.3
DES9	ChCl	LacA	1,3-propanediol	Water	1:5:1:0.91	6.8	14.9
DES10	ChCl	LacA	1,3-butanediol	Water	1:1:1:0.93	2.9	30.0
DES11	ChCl	LacA	1,3-butanediol	Water	1:2:1:0.92	4.5	22.9
DES12	ChCl	LacA	1,3-butanediol	Water	1:3:1:1	5.4	18.6
DES13	ChCl	LacA	1,3-butanediol	Water	1:4:1:1	6.1	16.7
DES14	ChCl	LacA	1,3-butanediol	Water	1:5:1:1	6.6	17.7
DES15	ChCl	LacA	1,4-butanediol	Water	1:1:1:0.96	3.0	30.1
DES16	ChCl	LacA	1,4-butanediol	Water	1:2:1:0.92	4.5	21.2
DES17	ChCl	LacA	1,4-butanediol	Water	1:3:1:0.92	5.5	18.8
DES18	ChCl	LacA	1,4-butanediol	Water	1:4:1:0.91	6.2	15.2
DES19	ChCl	LacA	1,4-butanediol	Water	1:5:1:0.91	6.7	14.4
DES20	ChCl	LacA	1,5-pentanediol	Water	1:1:1:0.87	3.9	29.8
DES21	ChCl	LacA	1,5-pentanediol	Water	1:2:1:0.98	5.2	22.3
DES22	ChCl	LacA	1,5-pentanediol	Water	1:3:1:0.90	5.9	19.5
DES23	ChCl	LacA	1,5-pentanediol	Water	1:4:1:0.90	6.7	18.0
DES24	ChCl	LacA	1,5-pentanediol	Water	1:5:1:0.96	6.9	15.1

Since all extractions were performed at 60 °C, the viscosity values at just 60 °C are given here. Complete viscosity data are gathered in Table 4.

Table 3. Characterization of different properties of DES (conductivity, refractive index, and density).

	Conductivity (mS/cm)	Refractive Index	Density (g/cm³)					
	25 °C	25 °C	25 °C	35 °C	45 °C	55 °C	65 °C	75 °C
DES1	1.87	1.4647	1.197	1.197	1.197	1.197	1.196	1.193
DES2	1.84	1.4562	1.099	1.099	1.099	1.099	1.099	1.099
DES3	1.76	1.4523	1.094	1.094	1.094	1.094	1.094	1.094
DES4	1.70	1.4499	1.070	1.070	1.070	1.070	1.069	1.068
DES5	3.45	1.4700	1.099	1.098	1.098	1.098	1.098	1.098
DES6	3.30	1.4614	1.078	1.078	1.078	1.078	1.078	1.077
DES7	2.99	1.4553	1.076	1.076	1.076	1.076	1.076	1.075
DES8	2.60	1.4516	1.063	1.063	1.063	1.063	1.063	1.062
DES9	2.28	1.4488	1.051	1.051	1.051	1.051	1.051	1.051
DES10	2.01	1.4689	1.083	1.083	1.082	1.082	1.082	1.082
DES11	1.95	1.4605	1.079	1.079	1.079	1.079	1.078	1.077
DES12	1.93	1.4547	1.073	1.073	1.073	1.073	1.073	1.073
DES13	1.76	1.4515	1.037	1.035	1.036	1.036	1.036	1.035
DES14	1.59	1.4484	1.029	1.023	1.028	1.028	1.028	1.027
DES15	2.44	1.4703	1.068	1.063	1.068	1.068	1.068	1.068
DES16	2.38	1.4619	1.067	1.067	1.067	1.067	1.067	1.067
DES17	2.27	1.4559	1.056	1.055	1.056	1.055	1.055	1.055
DES18	2.20	1.4527	1.053	1.053	1.053	1.053	1.053	1.051
DES19	2.08	1.4499	1.017	1.017	1.017	1.017	1.017	1.017
DES20	2.24	1.4689	1.080	1.080	1.080	1.080	1.080	1.079
DES21	2.14	1.4541	1.060	1.060	1.060	1.060	1.059	1.059
DES22	2.10	1.4539	1.058	1.058	1.058	1.058	1.057	1.057
DES23	1.96	1.4506	1.044	1.044	1.044	1.044	1.044	1.043
DES24	1.81	1.4500	1.037	1.037	1.037	1.037	1.037	1.036

Values of conductivity, refractive index and density were determined as described elsewhere [26,28].

2.3. Plant Materials

Spruce bark (*Picea abies*) as an industrial waste was provided by the timber company Bioenergo Ltd. (Ruzomberok, Slovakia). The spruce bark was air dried at ambient temperature until constant weight, homogenized by grinding using a knife mill with a motor power of 7.5 kW and separated using sieves into fractions. The 1.0–1.4mm fraction of spruce bark was extracted using DESs and analyzed to determine the content of holocellulose (52.0% ± 0.2%), lignin (26.4% ± 1.3%), ash (3.6% ± 0.4%) and extractives (12.7% ± 0.01%). The humidity of the material (8.77% ± 0.08%) was determined by drying approximately 1g of spruce bark at 105°C for 6 hours until complete water removal.

2.4. Extraction

The extraction conditions were similar to those described in Škulcová et al. [29]. Homogeneous samples of bark were withdrawn from the bark storage system in all extraction experiments. The dried and weighed ground bark was added to the DESs at a 1:20 (wt/wt) ratio. The extraction was performed for 2 h at 60 °C under continuous stirring in a closed flask.

When deciding on the application of DESs on an industrial scale, one of the key factors is their thermal stability. This stability is not just a function of thermal stability of their constituents but is also influenced by hydrogen bonds and electrostatic interactions, both decreasing with increases in temperature. High temperature might cause changes in the mass of DES due to its evaporation or decomposition. Haz et al. [62] investigated long-term isothermal stability of DESs (10 hours, 60–120 °C) composed of choline chloride and an organic acid (lactic, tartaric or malonic). Based on the results obtained it may be said that DESs investigated are stable at 60 °C. Lynam et al. [63] investigated in detail five DESs containing an organic acid (lactic, formic, acetic) and an amino acid (betaine, proline) or chloride choline. The thermal analysis took place in an inert nitrogen atmosphere at a heating rate 20 °C/min in the temperature range 30–100 °C, subsequently followed by heating rate 10 °C/min in 100–160 °C. The selected temperature limit 160 °C represented the maximum temperature of biomass processing. The authors compared boiling temperatures of the DESs respective constituents and thermal stability of the DESs. All DESs investigated were thermally stable up to 160 °C [63]. Based on the results above, it can be concluded that all the DESs used are thermally stable. The water content in the DESs was determined by coulometric Karl-Fischer titration, and during extraction in a closed flask it did not change.

2.5. Determination of Total Phenolic Content

The TPC in the extracts was estimated spectrometrically according to our previous work [29], based on redox reactions of Folin-Ciocalteu's reagent with phenols. First, 0.25 g of the extract was added into a 10 mL flask, and the flask was filled with ethanol. A total of 0.25 mL from the stock solution was mixed with 0.25 mL of Folin-Ciocalteu's reagent and 1.25 mL of 20% Na_2CO_3 p.a. solution in a 10 mL volumetric bank, which was then filled with distilled water. After agitation and standing for 1 h at an ambient temperature, the absorbance of the solution was measured against blanks in 0.5 cm cells at a wavelength of 765 nm. The phenolic compounds were expressed as gallic acid equivalent (GAE) in 100 g of extract using a calibration curve in the form of a straight line. All measurements were performed three times for each individual sample. The data in Table 5 represent average values; the differences in measurements did not exceed 3%.

2.6. Determination of Antioxidant Activity

The antioxidant activity was determined as free radical scavenging activity (RSA), using a standard method [64] based on the discoloration of the samples after reacting with the stable free 2,2-diphenyl-1-picrylhydrazyl radical (•DPPH), and subsequent absorbance measurements at 517 nm. Briefly, 3.5 mg/mL of extract was mixed with fresh •DPPH (0.08 mg/mL in methanol) solution at a ratio of 1:1 (vol/vol). The absorbance of the tested extracts, measured at 517 nm, was read against a blank (methanol) after 0, 5, 10, 15, 20, 25 and 30 min. Gallic acid was used as a reference and corresponded to 100% activity. The RSA was calculated using Equation (1):

$$RSA\ (\%) = 100 \times (A_0 - A_{TEST})/(A_0 - A_{REF}) \tag{1}$$

where A_0 is the initial absorbance of the •DPPH solution in methanol, A_{TEST} is the absorbance of the tested sample in the •DPPH solution, and A_{REF} is the absorbance of gallic acid (0.7 mg/mL in methanol) in the •DPPH solution.

2.7. Determination of Viscosity

The viscoelastic properties were evaluated using a Brookfield DVII + Pro viscometer, as described earlier [27,28]. The sample viscosity was measured at different temperatures (30–90 °C) and revolutions (5, 10, 20 and 50 rpm), using a spindle 18 with an adapter. All the measurements were performed three times on individual samples. The error of measurement for individual viscosities is in the range of 0.2 to 0.5 mPa·s. Temperature and rpm dependences of viscosity for the used DESs are listed in Table 4. The resulting viscosity for different temperatures is expressed as the average value for different revolutions. All systems behaved as Newtonian liquids.

Table 4. Viscosity of the studied DESs for different temperatures and revolutions (5–50 rpm).

Temperature (°C)	Viscosity (mPa s)				Average Viscosity (mPa·s)
	5 rpm	10 rpm	20 rpm	50 rpm	
DES1: ChCl:LacA:Water (1:2:0.96)					
30	133.3	134.4	134.1	x	134.1
40	79.2	78.6	78.1	x	78.6
50	49.2	47.4	47.2	48.1	48.0
60	30.0	31.5	31.6	31.3	31.1
70	22.2	22.8	22.3	22.3	22.4
80	16.2	16.5	16.8	16.5	16.5
90	13.2	12.9	13.8	13.5	13.4
DES2: ChCl:LacA:Water (1:3:0.97)					
30	94.8	96.0	96.7	x	95.8
40	58.2	57.6	56.7	57.2	57.4
50	36.3	35.1	35.7	35.5	35.6
60	25.2	26.4	26.2	26.5	26.1
70	17.4	18.9	18.6	18.4	18.3
80	15.0	13.8	13.9	14.1	14.2
90	11.4	11.4	10.8	10.6	11.1
DES3: ChCl:LacA:Water (1:4:0.99)					
30	78.6	79.2	79.2	x	79.0
40	49.8	47.7	47.2	48.0	48.2
50	29.0	30.9	30.7	30.3	30.2
60	21.6	21.6	21.0	21.1	21.3
70	16.8	15.0	15.6	15.4	15.7
80	12.6	12.6	12.6	12.6	12.6
90	9.6	9.6	9.9	10.0	9.8

Table 4. Cont.

Temperature (°C)	Viscosity (mPa s)				Average Viscosity (mPa·s)
DES4: ChCl:LacA:Water (1:5:0.98)					
	5 rpm	10 rpm	20 rpm	50 rpm	
30	73.2	71.7	71.4	x	72.1
40	42.6	41.7	42.1	42.5	42.2
50	26.4	27.3	27.1	27.1	27.0
60	19.0	19.2	18.6	18.8	18.9
70	12.6	14.2	14.2	13.7	13.7
80	11.1	11.4	11.1	11.1	11.2
90	8.4	8.7	8.9	9.1	8.8
DES5: ChCl:LacA:1.3-propanediol:Water (1:1:1:0.92)					
	5 rpm	10 rpm	20 rpm	50 rpm	
30	86.4	85.5	84.9	x	85.6
40	57.2	55.2	54.4	54.3	55.3
50	36.6	36.0	36.4	36.0	36.3
60	25.2	26.4	25.5	25.0	25.5
70	18.6	18.9	18.1	18.1	18.4
80	14.4	13.8	13.8	13.9	14.0
90	10.8	10.5	10.6	10.7	10.7
DES6: ChCl:LacA:1.3-propanediol:Water (1:2:1:0.95)					
	5 rpm	10 rpm	20 rpm	50 rpm	
30	65.4	64.2	63.6	x	64.4
40	40.2	38.7	40.0	40.1	39.8
50	26.4	27.0	25.8	26.2	26.4
60	18.6	17.7	18.3	18.1	18.2
70	13.2	12.9	13.5	13.7	13.3
80	10.8	10.2	10.2	10.3	10.4
90	8.2	8.1	8.4	8.4	8.3
DES7: ChCl:LacA:1.3-propanediol:Water (1:3:1:0.91)					
	5 rpm	10 rpm	20 rpm	50 rpm	
30	57.6	57.6	56.4	56.4	57.0
40	34.8	33.6	34.9	34.8	34.5
50	21.0	23.1	22.6	22.8	22.4
60	15.6	16.2	16.0	15.8	15.9
70	12.6	11.7	11.8	12.0	12.0
80	9.0	9.3	9.1	9.1	9.1
90	7.2	7.8	7.6	7.6	7.6
DES8: ChCl:LacA:1.3-propanediol:Water (1:4:1:0.92)					
	5 rpm	10 rpm	20 rpm	50 rpm	
30	51.6	50.1	49.8	50.5	50.5
40	31.2	31.2	31.3	31.0	31.2
50	21.6	22.2	21.7	22.1	21.9
60	15.3	15.0	15.6	15.4	15.3
70	12.6	11.4	11.7	11.6	11.8
80	8.4	9.0	9.0	9.1	8.9
90	6.6	7.5	7.4	7.3	7.2

Table 4. Cont.

Temperature (°C)	DES9: ChCl:LacA:1.3-propanediol:Water (1:5:1:0.91)				
	Viscosity (mPa s)				Average Viscosity (mPa·s)
	5 rpm	10 rpm	20 rpm	50 rpm	
30	46.2	44.4	44.2	44.6	44.9
40	28.3	28.2	28.2	27.9	28.3
50	19.3	21.6	20.4	20.2	20.5
60	15.5	14.4	14.8	14.7	14.9
70	12.0	11.4	11.1	10.7	11.3
80	11.4	10.8	10.8	10.5	10.9
90	9.6	9.6	8.9	8.3	9.2

Temperature (°C)	DES10: ChCl:LacA:1.3-butanediol:Water (1:1:1:0.93)				
	Viscosity (mPa s)				Average Viscosity (mPa·s)
	5 rpm	10 rpm	20 rpm	50 rpm	
30	113.2	117.9	118.5	x	x
40	72.0	71.4	71.1	x	x
50	46.8	45.3	44.7	45.4	45.4
60	29.4	29.7	30.4	30.0	30.0
70	21.6	21.9	20.8	21.1	21.1
80	16.2	15.6	15.9	15.8	15.8
90	12.6	12.6	12.1	12.1	12.1

Temperature (°C)	DES11: ChCl:LacA:1.3-butanediol:Water (1:2:1:0.92)				
	Viscosity (mPa s)				Average Viscosity (mPa·s)
	5 rpm	10 rpm	20 rpm	50 rpm	
30	86.4	86.1	85.8	x	86.1
40	52.8	51.9	51.1	51.9	51.9
50	33.0	32.7	33.1	32.8	32.9
60	22.8	22.5	21.9	22.2	22.4
70	16.2	15.6	16.0	15.7	15.9
80	12.0	12.0	11.8	11.9	11.9
90	9.5	10.5	9.9	10.1	10.0

Temperature (°C)	DES12: ChCl:LacA:1.3-butanediol:Water (1:3:1:1)				
	Viscosity (mPa s)				Average Viscosity (mPa·s)
	5 rpm	10 rpm	20 rpm	50 rpm	
30	72.6	71.7	71.4	x	71.9
40	43.2	42.0	42.7	42.8	42.7
50	25.8	27.3	27.4	27.2	26.9
60	19.2	18.6	18.1	18.3	18.6
70	13.8	12.3	13.5	13.3	13.2
80	10.2	10.5	10.2	10.3	10.3
90	7.8	9.0	8.8	9.0	8.7

Temperature (°C)	DES13: ChCl:LacA:1.3-butanediol:Water (1:4:1:1)				
	Viscosity (mPa s)				Average Viscosity (mPa·s)
	5 rpm	10 rpm	20 rpm	50 rpm	
30	63.6	63.0	62.4	x	63.0
40	38.4	38.1	38.5	38.5	38.4
50	23.4	24.9	24.3	24.5	24.3
60	17.4	16.2	16.8	16.5	16.7
70	12.0	12.3	12.1	12.1	12.1
80	9.6	9.6	9.4	9.4	9.5
90	6.6	7.2	7.1	7.4	7.1

Table 4. *Cont.*

Temperature (°C)	Viscosity (mPa s)				Average Viscosity (mPa·s)
DES14: ChCl:LacA:1.3-butanediol:Water (1:5:1:1)					
	5 rpm	10 rpm	20 rpm	50 rpm	
30	63.6	60.9	60	59.9	61.1
40	38.4	36.7	36.3	35.9	36.8
50	24.6	26.1	24.0	23.7	24.6
60	19.8	17.7	16.8	16.6	17.7
70	13.8	12.6	12.7	12.3	12.9
80	10.2	10.5	9.0	9.1	9.7
90	8.4	8.4	8.4	8.5	8.4
DES15: ChCl:LacA:1.4-butanediol:Water (1:1:1:0.96)					
	5 rpm	10 rpm	20 rpm	50 rpm	
30	107.4	105.9	106.9	x	106.7
40	69.6	68.7	68.2	x	68.8
50	45.0	43.2	43.8	44.1	44.0
60	30.0	30.3	30.1	30.0	30.1
70	21.6	21.9	21.1	21.3	21.5
80	16.0	15.6	16.0	15.9	15.9
90	12.0	12.9	11.8	11.9	12.2
DES16: ChCl:LacA:1.4-butanediol:Water (1:2:1:0.92)					
	5 rpm	10 rpm	20 rpm	50 rpm	
30	77.4	77.1	76.8	x	77.1
40	48.0	46.5	46.5	47.1	47.0
50	30.6	30.0	30.6	30.5	30.4
60	21.3	21.3	21.0	21.1	21.2
70	16.2	14.7	15.4	15.2	15.4
80	13.2	12.0	12.3	12.2	12.4
90	9.0	9.9	9.3	9.7	9.5
DES17: ChCl:LacA:1.4-butanediol:Water (1:3:1:0.92)					
	5 rpm	10 rpm	20 rpm	50 rpm	
30	64.8	63.3	62.7	x	63.6
40	39.6	38.4	39.7	39.7	39.4
50	25.2	27.3	26.1	26.5	26.3
60	19.2	18.9	18.7	18.2	18.8
70	14.4	14.1	14.4	14.2	14.3
80	10.8	11.1	11.4	11.3	11.2
90	9.6	9.6	9.5	9.5	9.6
DES18: ChCl:LacA:1.4-butanediol:Water (1:4:1:0.91)					
	5 rpm	10 rpm	20 rpm	50 rpm	
30	57.0	56.1	54.6	55.2	55.7
40	34.2	33.0	34.0	33.7	33.7
50	21.6	23.1	22.0	22.1	22.2
60	15.0	15.0	15.6	15.3	15.2
70	12.6	12.0	11.8	11.5	12.0
80	9.6	9.9	9.0	9.2	9.4
90	7.5	7.5	7.4	7.9	7.6

Table 4. Cont.

Temperature (°C)	DES19: ChCl:LacA:1.4-butanediol:Water (1:5:1:0.91)				
	Viscosity (mPa s)				Average Viscosity (mPa·s)
	5 rpm	10 rpm	20 rpm	50 rpm	
30	52.8	50.7	50.1	50.5	51.0
40	31.2	31.5	31.5	31.3	31.4
50	21.0	21.3	20.2	20.3	20.7
60	14.4	14.4	14.5	14.2	14.4
70	10.8	10.5	10.4	10.3	10.5
80	7.2	8.1	8.1	8.2	7.9
90	6.6	6.6	6.5	6.7	6.6

Temperature (°C)	DES20: ChCl:LacA:1.5-pentanediol:Water (1:1:1:0.87)				
	Viscosity (mPa s)				Average Viscosity (mPa·s)
	5 rpm	10 rpm	20 rpm	50 rpm	
30	107.4	108.6	109.3	x	108.4
40	69.0	68.7	68.4	x	68.7
50	45.0	43.0	43.9	44.3	44.1
60	28.8	30.0	30.4	30.1	29.8
70	21.0	21.6	21.3	21.5	21.4
80	16.2	15.9	16.2	15.0	16.1
90	12.6	12.3	12.6	12.4	12.5

Temperature (°C)	DES21: ChCl:LacA:1.5-pentanediol:Water (1:2:1:0.98)				
	Viscosity (mPa s)				Average Viscosity (mPa·s)
	5 rpm	10 rpm	20 rpm	50 rpm	
30	81.6	79.2	79.3	x	80.0
40	51.6	48.6	48.3	49.1	49.4
50	31.2	31.5	31.8	31.7	31.6
60	21.6	22.8	22.2	22.4	22.3
70	15.0	15.5	16.5	15.9	15.7
80	12.0	12.3	12.4	12.2	12.2
90	9.0	10.2	9.3	9.7	9.6

Temperature (°C)	DES22: ChCl:LacA:1.5-pentanediol:Water (1:3:1:0.90)				
	Viscosity (mPa s)				Average Viscosity (mPa·s)
	5 rpm	10 rpm	20 rpm	50 rpm	
30	76.2	74.7	74.2	x	75.0
40	47.4	44.7	44.5	44.9	45.4
50	28.8	29.4	29.1	28.7	29.0
60	19.2	19.8	19.3	19.5	19.5
70	14.4	13.8	14.5	14.1	14.2
80	10.8	11.1	10.6	10.5	10.8
90	7.8	8.7	7.9	8.3	8.2

Temperature (°C)	DES23: ChCl:LacA:1.5-pentanediol:Water (1:4:1:0.90)				
	Viscosity (mPa s)				Average Viscosity (mPa·s)
	5 rpm	10 rpm	20 rpm	50 rpm	
30	61.8	61.8	61.6	x	61.7
40	37.2	37.2	37.9	37.9	37.6
50	25.2	26.1	25.6	25.9	25.7
60	18.6	17.7	18.0	17.7	18.0
70	13.2	13.2	12.9	12.9	13.1
80	10.8	10.5	9.8	9.8	10.2
90	8.4	8.4	8.4	8.4	8.4

Table 4. *Cont.*

Temperature (°C)	Viscosity (mPa s)				Average Viscosity (mPa·s)
DES24: ChCl:LacA:1.5-pentanediol:Water (1:5:1:0.96)					
	5 rpm	10 rpm	20 rpm	50 rpm	
30	55.8	54.9	53.8	54.9	54.9
40	32.7	32.7	33.6	33.3	33.1
50	21.6	22.5	21.4	21.8	21.8
60	15.0	14.8	15.2	15.4	15.1
70	10.8	11.1	11.2	11.2	11.1
80	7.8	9.0	8.8	9.0	8.7
90	6.6	6.9	7.1	7.0	6.9

3. Results and Discussion

The extraction of phenolic compounds using DES systems has been performed and the results have been evaluated. Water is the most common solvent. It is able to form hydrogen bonds and plays the role of both hydrogen bond donor and acceptor [65,66]. Addition of water into DESs in the process of their formation causes incorporation of water molecules into the structure of DESs, their fixation by hydrogen bonds, and this water cannot be later fully removed by, e.g., a rotary evaporator. Different content of water in the studied DESs indicates that water molecules bind predominantly to hydrogen bond donors [66]. A small addition of water may result in a decrease in viscosity, temperature lowering and shorter time needed for DES preparation. Moreover, the presence of water may change the ability to dissolve some compounds in DES, e.g., plant metabolites [16]. Water influences the course of reactions where a DES acts as a catalyst promoter. Understanding how water activity changes the DESs properties is key for researchers looking to encourage or retard biological growth in addition to studies of proton-coupled electron transfer, solvation, or any other water-coupled or water-dependent process [67]. Smith et al. [67] examined the influence of water in the system of choline chloride and urea (1:2) which was published in a paper in 2019. Based on their results it is possible to draw a conclusion that if DES systems contain water, it is necessary to take water into account as another component of the DES. Therefore, binary systems need to be characterized as ternary.

In this work, water containing choline chloride-based deep eutectic solvents (DESs) with lactic acid and 1,3-propanediol, 1,3-butanediol, 1,4-butanediol, and 1,5-pentanediol, with different molar ratios (Table 2), were used as extractants. Twenty-four different DESs (Table 1) were applied as extractants. As shown in Table 5, the content of polyphenols in the eutectic extracts ranged from 177.6 to 596.2 mg of gallic acid equivalent (GAE) per 100 g of dry bark.

The amount of total phenolic compounds varied widely in different plant materials (92 phenolic extracts from acetone and methanol were examined) and ranged from 0.2 to 155.3 mg GAE/g dry material [68]. The yields of phenolic compounds were generally variable depending on the tree species, part of the tree used for extraction, as well as the age or place of origin of the tree [69,70]. According to the literature, in forest pine, the total amount of polyphenols ranges between 76 mg GAE /g in dry bark, 17.5 mg GAE/g in needles, and 1.1 mg GAE/g in cork wood [48]. The concentration of phenolic compounds in spruce knotwood is 10–15%, and even up to 30%, of absolute dry weight, while in pine knotwood it is less than 10% dry weight, with concentrations several times lower than that observed for logs [71]. According to [71], more than half of the hydrophilic extractive substances in lignite are found in knotwood.

In the previous paragraph, the extractions using organic solvents are presented. The next paragraph deals with our experiments applying DESs.

Table 5. Total phenolic contents of the spruce bark extracts obtained with DESs.

Sample	TPC (mg GAE/100 g Extract)	TPC (mg GAE/100 g Dry Bark)
DES1	15.7 ± 0.1	393.6 ± 3.0
DES2	13.9 ± 0.3	336.9 ± 6.5
DES3	12.8 ± 0.2	326.7 ± 3.0
DES4	13.4 ± 0.1	349.4 ± 2.6
DES5	11.6 ± 0.2	288.1 ± 2.5
DES6	13.8 ± 0.4	336.5 ± 8.7
DES7	12.0 ± 0.3	312.3 ± 7.3
DES8	14.0 ± 0.1	361.6 ± 2.2
DES9	13.8 ± 0.1	343.8 ± 2.1
DES10	9.4 ± 0.1	233.6 ± 1.8
DES11	10.9 ± 0.4	287.6 ± 8.5
DES12	11.6 ± 0.1	277.6 ± 2.1
DES13	20.9 ± 0.2	531.4 ± 3.9
DES14	23.4 ± 0.3	596.2 ± 7.4
DES15	12.1 ± 0.1	283.6 ± 2.0
DES16	13.4 ± 0.1	331.7 ± 2.5
DES17	12.3 ± 0.3	313.3 ± 7.4
DES18	13.7 ± 0.2	337.9 ± 3.9
DES19	12.6 ± 0.1	339.1 ± 1.8
DES20	13.7 ± 0.1	332.6 ± 2.1
DES21	14.3 ± 0.1	350.4 ± 3.0
DES22	14.6 ± 0.2	363.9 ± 2.8
DES23	11.3 ± 0.2	291.8 ± 5.1
DES24	16.0 ± 0.1	422.4 ± 2.4

A series of experiments applying DESs was performed to examine the effect of lactic acid as a hydrogen bond donor on the efficiency of polyphenol extraction. In the series DES1 to DES4, the choline chloride: lactic acid: water molar ratio was varied from 1:2:0.96 to 1:5:0.98. The results showed that the highest polyphenol content was measured for the extract with the 1:2:0.96 molar ratio (396.6 mg GAE/100 g dry bark). Thus, increasing the HBD (lactic acid) content in the extractant did not cause an increment in the TPC. In contrast, at the molar ratio of 1:3:0.97, the polyphenol content was 336.9 mg GAE/100 g dry bark, and at 1:4:0.99 – only 326.7 mg GAE/100 g dry bark. At the molar ratio of 1:5:0.98, the TPC increased to 349.4 mg GAE/100 g dry bark.

It is known that the recovery of polar compounds from samples can be enhanced varying the extraction conditions (diffusivity, density, viscosity) and with the addition of a co-solvent [53]. As shown in Table 1, viscosity decreases with increasing HBD content. Thus, we assumed that, under the same extraction conditions, the TPC should increase with a decrease in viscosity. However, this has not been confirmed.

As mentioned earlier, the addition of a co-solvent also affects the amount of substances obtained. The addition of a co-solvent (e.g., ethanol) is intended to significantly swell plant cells, allowing better solvent penetration and diffusion of the solute present in the solid lignocellulosic matrix [72]. Similar behavior has been demonstrated in several previous studies [73–75]. However, the use of a higher concentration of the co-solvent can also reduce the yield of bioactive compounds because of CO_2-co-solvent interactions [59].

In our case, we added another HBD to the system, namely 1,3-propanediol, 1,3-butanediol, 1,4-butanediol, or 1,5-pentanediol. The aim was to reduce the viscosity and thus improve the extraction process. As a result, the viscosity of these systems was reduced at 60 °C, and therefore we assumed that the TPC would increase when applying these extraction systems. We found that in the DES5 to DES9 systems (i.e., ChCl : LacA : 1,3-propanediol : water (molar ratios 1 : 1 : 1 : 0.92; 1 : 2 : 1 : 0.95; 1 : 3 : 1 : 0.91; 1 : 4 : 1 : 0.92; 1 : 5 : 1 : 0.91), the TPC reached 288.1; 336.5; 312.3; 361.3 and 343.8 mg

GAE/100 g dry bark, respectively. Thus, the addition of 1,3-propanediol had no significant effect on the TPC and, in some cases, there was even a decrease in the TPC. In the case of the system containing 1,3-butanediol, the TPC ranged from 277.6 to 562.2 mg GAE/100 g dry bark. When using 1,4-butanediol, the TPC reached 283.6–337.9 mg GAE/100 g dry bark. Thus, the changes in the extraction system and viscosity (viscosity at 60 °C ranged from 30.1 mPa s to 14.4 mPa s) altered the TPC. The addition of 1,5-pentanediol resulted in TPC ranging from 291.8 to 422.4 mg GAE/100 g dry bark. The highest TPC was achieved by the system having a 1 : 5 : 1 : 0.96 molar ratio (422.4 mg GAE/100 g dry bark). For the 1 : 3 : 1 : 0.9 system, the TPC increased by 8% over the ChCl : LacA : Water system (1 : 3 : 0.97). For other extraction systems, a decrease in the TPC was achieved, despite the reduced viscosity, compared to the diol-free systems (DES1, DES3).

Based on this evaluation of the investigated 24 extraction systems, we can conclude that the system containing ChCl : LacA : 1,3-butanediol : water (1 : 4 : 1 : 1; 1 : 5 : 1 : 1) achieved the best results in polyphenols extraction. Both systems displayed a significant increase in the TPC for the 1 : 4 : 1 : 1 system, the TPC was 531.4 mg GAE/100 g dry bark, while for the 1 : 5 : 1 : 1 molar ratio, the TPC was up to 596.2 mg GAE/100 g dry bark (23.4 mg GAE/100 g extract).

As mentioned in the Introduction, the extraction of polyphenols from various wastes is a very important area of research. Nevertheless, to the best of the authors' knowledge, only one paper has been published on the extraction of spruce bark using DESs [29]. Using 41 DESs, the polyphenols content was reported to range from 41 to 463 mg GAE/100 g of extract (or 9 to 100 mg GAE per 1 g dry weight [29]).

In this work, the TPC content achieved in the extract ranged from 9.4 to 23.4 mg GAE/100 g extract (177.6 to 596.2 mg GAE/100 g of dry bark). From the viewpoint of implementing DESs in practice, it is necessary to compare the TPC reached using DESs with that achieved using common organic solvents (taking into account the impact of bark diversity). Working with ethanol-water mixtures (50%–70% v/v ethanol) at 40–60 °C and extraction time of 30–60 minutes, applying UAE, the TPC was in the range of 6.62–13.32 mg GAE/g spruce bark [10]. The TPC in the spruce bark extract (*Picea abies* L.) using the classical method (water batch extraction) was 113.56 mg GAE/g extract, while when the sonication method was used, the TPC value was lower (84.28 mg GAE/g extract) [76]. Spinelli et al. [7], published the results obtained using supercritical fluid extraction (SFE), pressurized liquid extraction (PLE), and ultrasound-assisted extraction (UAE) of Norway spruce bark by ethanol. The TPC for SFE and different ethanol concentrations were as follows: 10% ethanol: 0.77 mg GAE/g DM; 20% ethanol: 1.24 mg GAE/g DM; and 40% ethanol: 2.5 mg GAE/g DM. For PLE and water, 33.45 mg GAE/g DM was determined; for 98% ethanol, 46.34 mg GAE/g DM was measured. Using UAE and 98% ethanol, 54.97 mg GAE/g DM was found.

In addition to the solvent type, the polyphenol content also depends largely on the bark particle size. Sládkova et al. [77] found that using MAE, the TPC varied between 42.7 and 265.0 mg GAE per 100 g of dry bark for different particle sizes (0.3; 1 and 2.5 mm) at an extraction temperature of 60 °C. When the particle size was 1 mm, the TPC ranged from 90.3 to 321.1 mg GAE/100 g dry bark. It is known that, normally, the amount of extracts will depend on many factors (particle size, method used, extracting agent, extraction conditions, and others) and a simple comparison is thus inadequate, being rather indicative. In this investigation, it has been shown that by changing the reagent through the addition of a third component, in some cases, it is possible to obtain a higher TPC. This is particularly related to a change in the polarity of the extraction system due to the addition of diols. In our work, the extraction system was very simple, a DES was mixed with the spruce bark at 60 °C and the mixture was extracted for 2 hours. As suggested by other authors regarding the treatment of various kinds of biomass using MAE and UAE, the yield of extracted polyphenols can be higher than that using conventional reagents [78–81].

A separate part of the work focused on the determination of the antioxidant capacity of the extracts. Radical scavenging activity of phenolic compounds is stemming from their ability to act as reducing agents, hydrogen or electrons donors, and singlet oxygen quenchers [82]. Table 6 summarizes the antioxidant activity measured at 0–30 min after the addition of •DPPH. The differences in radical scavenging activity (RSA) suggested that each DES preferentially dissolved another type of extractive with a differing reactivity to •DPPH. Moreover, each DES had a different extraction of TPC, and thus, the amount of extractives had an impact on the antioxidant activity and on the reaction with the radical. To clear up the matter, extractives are individual compounds reacting with •DPPH. RSA values relate to extracts (DES components and extractives) and were found to vary from 81.4% to 95%. A lower antioxidant activity was observed for the extracts obtained with ChCl : LacA : Water (1 :2 : 0.96). RSA 86.4%, and for the system containing ChCl : LacA and different diols in a molar ratio of 1 : 1 : 1, namely 82.4% for 1,3-propanediol; 84.2% for 1,3-butanediol; 85.4% for 1,4-butanediol, and 81.4% for 1,5-pentanediol. The ChCl : LacA : 1,3-butanediol : Water (1 : 5 : 1 : 1) extracts had the highest antioxidant activity, with RSA of 95% and this extract also contained the highest content of polyphenols (596.2 mg GAE/100 g dry bark).

In several works previously reported by other authors, focus is laid on finding a correlation between TPC and antioxidant activity [29,40,82,83]. In our study, we also tried to find such a correlation, however, with no success (figure not shown), the correlation coefficient (R^2) was just 0.12.

Table 6. 2,2-diphenyl-1-picrylhydrazyl radical (•DPPH) assay of antioxidant activity (RSA) for different extracts.

Sample	RSA (%)						
Time (min)	0	5	10	15	20	25	30
DES1	76.2	78.6	81.0	82.7	84.3	85.4	86.4
DES2	83.3	86.2	88.2	89.8	90.9	92.2	93.2
DES3	85.9	88.0	89.5	90.8	91.7	92.5	93.2
DES4	83.7	86.2	88.1	89.6	90.6	91.5	92.3
DES5	72.0	74.6	76.8	78.5	79.9	81.3	82.4
DES6	81.4	84.3	86.2	87.9	89.3	90.4	91.3
DES7	85.2	87.9	89.7	91.3	91.7	93.1	93.8
DES8	86.2	88.7	90.6	91.9	92.8	93.4	94.1
DES9	86.6	89.4	91.1	92.5	93.4	94.0	94.6
DES10	69.0	74.7	78.0	80.2	81.8	83.1	84.2
DES11	70.6	76.6	80.0	82.3	84.3	85.6	86.9
DES12	74.7	81.7	85.0	87.6	89.4	90.5	91.7
DES13	75.7	82.7	86.4	89.0	90.6	91.8	92.6
DES14	80.8	87.9	91.0	92.8	93.8	94.5	95.0
DES15	70.0	75.5	78.8	81.1	82.8	84.3	85.4
DES16	76.6	82.9	86.3	88.5	90.2	91.3	92.3
DES17	76.0	82.5	86.0	88.3	89.8	91.1	92.1
DES18	78.7	85.1	88.5	90.6	92.1	93.1	93.8
DES19	77.1	84.3	87.5	91.2	92.8	93.4	94.1
DES20	71.5	74.0	75.7	77.5	78.9	80.1	81.4
DES21	78.9	81.8	83.8	85.4	86.9	88.0	88.8
DES22	79.4	82.2	84.3	86.0	87.5	88.5	89.7
DES23	82.6	89.0	92.0	93.4	94.2	94.7	94.9
DES24	74.3	80.1	83.5	85.8	87.5	88.8	89.8

The recovery of polyphenolic substances from agri-food and forestry wastes is an important target in the future of biorefining and numerous studies have been oriented towards the valorization of processing wastes by the exploitation of side streams, implementing eco-friendly and cost-effective technologies [1–3,13,16–18,79–81,84–90].

The zero-waste biorefinery concept was supported by the work of Li et al. [87], who studied 12 refined oils, including grapeseed, rapeseed, peanut, sunflower, olive, avocado, almond, apricot, corn, wheat germ, soybean, and hazelnut oils to extract substances from rosemary leaves. The zero-waste concept deserves attention from the perspective of using DESs as well. One of the aims of using DESs may be to obtain an enriched solvent, without requiring any further process to separate the extracted substances. This enriched solvent would be used as syrup, either as a nutritional supplement with antioxidants, suitable for direct consumption, or as a preparation for skin, hair, and so on. treatment. Thus, this extraction technology would reduce the amount of waste generated, but would also be a cost-effective method for the isolation of value-added substances from forestry or agri-food residues. Moreover, the recycling process would be unnecessary, since the waste resulting after the extraction, especially in the case of agri-food waste, could be further used as animal feed by selecting the appropriate type of solvents for the extraction.

As far as the properties of choline chloride are concerned, it is characterized as a harmless compound, furthermore, it is an essential nutrient [82] and is used in the treatment of several diseases. As far as the price of choline chloride is concerned, high-quality 99% food grade or 99% pharmaceutical grade choline chloride is supplied for 10–30 $/kg; industrial 98% choline chloride is sold for 5–9 $/kg at minimal order of 20 kg (data available from www.alibaba.com). It thus can be considered as a cheap chemical. Similarly, lactic acid is low-cost, its price being of 1.1–2 $/kg, while that of ethanol is 0.5–1.5 $/kg.

When advocating the practical advantages of DESs, one of the most important supporting arguments is their ability of recuperation and multiple use [13]. Generally, DES recycling lies in the use of an anti-solvent, causing the elimination of a component from the system under operation, and after removal of water by evaporation, purified DES is obtained and may be reused [13].

On the other hand, it has been confirmed that, when certain types of DESs are applied, they achieve greater extractability of polyphenols than conventional solvents [78,89,90]. The antioxidant capacity achieved is also higher than that reached using conventional solvents [79–81]. It also appears that the extraction time at the same yield of extracted substances is lower when using DESs. In addition, it has been found that these solvents can be combined with other techniques, such as microwave-assisted or ultrasound-assisted extraction, which speed up the extraction time, extractive yield and, at the same time, use less solvent for the extraction [79–81].

4. Conclusions

This study focused on the application of deep eutectic solvents in spruce bark extraction. Twenty-four solvents based on combinations of choline chloride with lactic acid and 1,3-propanediol, 1,5-pentanediol, 1,4-butanediol, 1,3-butanediol, and water, with different molar ratios, were used as extractants under conditions of 120 min extraction time and a temperature of 60 °C. The total phenolic content and antioxidant activity of the extracts were determined. The results from the TPC analysis indicated that the extract achieved with choline chloride, lactic acid, 1,3-butanediol and water (1 : 5 : 1 : 1) had the highest antioxidant activity and radical scavenging activity of 95%. Also, this extract contained the highest content of polyphenols (596.2 mg GAE/100 g dry bark).

A comparison of the TPC and RSA values obtained in bark extraction using DESs under ambient conditions with those reached with organic solvents at higher temperature and pressure shows that the organic solvents are, in most cases, more effective. A comparison of the total cost associated with the use of the two mentioned classes of extractants may, however, lead to contrasting results. The reason lies in the cost of handling, cleaning and recovering organic solvents, the costs of generating and maintaining high temperature and pressure, as well as the costs of maintaining the quality of the living and working environment. It can be assumed that, over time, the factors mentioned will favor the use of DESs over that of organic solvents.

When evaluating the use of antioxidants extracted from biomass, two new aspects are emerging. The first is that, along with the traditional applications of antioxidants of natural origin in medicine, pharmacy, cosmetics, and food industry, they might be introduced into other fields as well. The following areas can be suggested as examples: the stabilization of dyes in the textile industry [91,92]; the stabilization of biofuels [93,94]; the stabilization of polymers [95,96]; and metalworking [97].

Of course, any application requires knowledge of the mechanism of action of antioxidants [98]. The second aspect is related to the form of use of antioxidants. It would be very advantageous if extracts could be used instead of isolated individual antioxidants. In this regard, the use of DESs for the extraction of antioxidants seems to be particularly convenient.

Author Contributions: M.J. and J.S. contributed equally to the conceptualization and design of the work; writing—original draft preparation: M.J., formal analysis: P.S. and V.M.; investigation: J.J., P.S., and V.M.; writing—editing: M.J., J.S., and J.J.; supervision and critical revision of the manuscript: M.J., J.S.; project administration: M.J.; funding acquisition: M.J. All of the authors approved the version of the manuscript submitted for publication. All authors have read and agreed to the published version of the manuscript.

Funding: This work was supported by the Slovak Research and Development Agency under the contracts Nos. APVV-15-0052, APVV-19-0185, APVV-16-0088, APVV-19-0174 and VEGA 1/0403/19. This article was also realized thanks to the support for infrastructure equipment provided by the Operational Program "Research and Development" for the project "National Center for Research and Application of Renewable Energy Sources" (ITMS 26240120016, ITMS 26240120028), for the project "Competence Center for New Materials, Advanced Technologies and Energy" (ITMS 26240220073), and for the project "University Science Park STU Bratislava" (ITMS 26240220084), co-financed by the European Regional Development Fund.

Acknowledgments: The authors would like to acknowledge the financial support of the Slovak Research and Development Agency and the support for infrastructure equipment provided by the Operational Program "Research and Development".

Conflicts of Interest: The authors declare no conflict of interest. The funders had no role in the design of the study; in the collection, analyses, or interpretation of data; in the writing of the manuscript, or in the decision to publish the results.

Nomenclature

RSA—radical scavenging activity (%)TPC—total phenolic content (mg GAE/100 g dry bark)

Abbreviations

ABTS—2,20-azinobis (3-ethylbenzothiazoline-6-sulfonic acid)
ASE—accelerated solvent extraction
ChCl—choline chloride
DES—deep eutectic solvent
DM—dry material
DPPH—2,2-diphenyl-1-picrylhydrazyl radical
FRAP—ferric reducing antioxidant power
GAE—gallic acid equivalents
HBA—hydrogen bond acceptor
HBD—hydrogen bond donor
LacA—lactic acid
MAE—microwave-assisted extraction
NADES—natural deep eutectic solvent
PLE—pressurized liquid extraction
SFE—supercritical fluid extraction
TEs—Trolox equivalent
UAE—ultrasound-assisted extraction

References

1. Jablonský, M.; Škulcová, A.; Malvis, A.; Šima, J. Extraction of value-added components from food industry based and agro-forest biowastes by deep eutectic solvents. *J. Biotechnol.* **2018**, *282*, 46–66. [CrossRef] [PubMed]
2. Mbous, Y.P.; Hayyan, M.; Hayyan, A.; Wong, W.F.; Hashim, M.A.; Looi, C.Y. Applications of deep eutectic solvents in biotechnology and bioengineering—Promises and challenges. *Biotechnol. Adv.* **2017**, *35*, 105–134. [CrossRef] [PubMed]
3. Jablonský, M.; Nosalova, J.; Sladkova, A.; Ház, A.; Kreps, F.; Valka, J.; Miertus, S.; Frecer, V.; Ondrejovic, M.; Sima, J.; et al. Valorisation of softwood bark through extraction of utilizable chemicals. A review. *Biotechnol. Adv.* **2017**, *35*, 726–750. [CrossRef] [PubMed]
4. Smith, E.L.; Abbott, A.P.; Ryder, K.S. Deep Eutectic Solvents (DESs) and Their Applications. *Chem. Rev.* **2014**, *114*, 11060–11082. [CrossRef]

5. Watson, R.R. *Polyphenols in Plants: Isolation, Purification and Extract Preparation*; Academic Press: New York, NY, USA, 2018.
6. Bianchi, S.; Koch, G.; Janzon, R.; Mayer, I.; Saake, B.; Pichelin, F. Hot water extraction of Norway spruce (*Picea abies* [*Karst.*]) bark: Analyses of the influence of bark aging and process parameters on the extract composition. *Holzforschung* **2016**, *70*, 619–631. [CrossRef]
7. Spinelli, S.; Costa, C.; Conte, A.; La Porta, N.; Padalino, L.; Del Nobile, M.A. Bioactive Compounds from Norway Spruce Bark: Comparison Among Sustainable Extraction Techniques for Potential Food Applications. *Foods* **2019**, *8*, 524. [CrossRef]
8. Krogell, J.; Holmbom, B.; Pranovich, A.; Hemming, J.; Willför, S. Extraction and chemical characterization of Norway spruce inner and outer bark. *Nord. Pulp Pap. Res. J.* **2012**, *27*, 6–17. [CrossRef]
9. Co, M.; Fagerlund, A.; Engman, L.; Sunnerheim, K.; Sjöberg, P.J.R.; Turner, C. Extraction of antioxidants from spruce (*Picea abies*) bark using eco-friendly solvents. *Phytochem. Anal.* **2011**, *23*, 1–11. [CrossRef]
10. Ghitescu, R.-E.; Volf, I.; Carausu, C.; Bühlmann, A.-M.; Gilca, I.A.; Popa, V.I. Optimization of ultrasound-assisted extraction of polyphenols from spruce wood bark. *Ultrason. Sonochem.* **2015**, *22*, 535–541. [CrossRef]
11. Krishnaiah, D.; Sarbatly, R.; Nithyanandam, R. A review of the antioxidant potential of medicinal plant species. *Food Bioprod. Process.* **2011**, *89*, 217–233. [CrossRef]
12. Kasote, D.; Katyare, S.S.; Hegde, M.V.; Bae, H. Significance of Antioxidant Potential of Plants and its Relevance to Therapeutic Applications. *Int. J. Boil. Sci.* **2015**, *11*, 982–991. [CrossRef] [PubMed]
13. Jablonský, M.; Šima, J. *Deep Eutectic Solvents in Biomass Valorization*; Spektrum STU: Bratislava, Slovakia, 2019; p. 176.
14. Santos-Sánchez, N.-F.; Salas-Coronado, R.; Villanueva-Cañongo, C.; Hernández-Carlos, B. Antioxidant Compounds and Their Antioxidant Mechanism. *Antioxidants* **2019**. [CrossRef]
15. Pisoschi, A.M.; Negulescu, G.P. Methods for Total Antioxidant Activity Determination: A Review. *Biochem & Anal Biochem* **2011**, *1*, 106. [CrossRef]
16. Dai, Y.; Van Spronsen, J.; Witkamp, G.-J.; Verpoorte, R.; Choi, Y.H. Natural deep eutectic solvents as new potential media for green technology. *Anal. Chim. Acta* **2013**, *766*, 61–68. [CrossRef]
17. Ekezie, F.-G.C.; Sun, D.-W.; Cheng, J.-H. Acceleration of microwave-assisted extraction processes of food components by integrating technologies and applying emerging solvents: A review of latest developments. *Trends Food Sci. Technol.* **2017**, *67*, 160–172. [CrossRef]
18. Zainal-Abidin, M.H.; Hayyan, M.; Hayyan, A.; Jayakumar, N.S. New horizons in the extraction of bioactive compounds using deep eutectic solvents: A review. *Anal. Chim. Acta* **2017**, *979*, 1–23. [CrossRef]
19. Shishov, A.; Bulatov, A.; Locatelli, M.; Carradori, S.; Andruch, V. Application of deep eutectic solvents in analytical chemistry. A review. *Microchem. J.* **2017**, *135*, 33–38. [CrossRef]
20. Ruesgas-Ramón, M.; Figueroa-Espinoza, M.C.; Durand, E. Application of Deep Eutectic Solvents (DES) for Phenolic Compounds Extraction: Overview, Challenges, and Opportunities. *J. Agric. Food Chem.* **2017**, *65*, 3591–3601. [CrossRef]
21. Aydin, F.; Yılmaz, E.; Soylak, M. Vortex assisted deep eutectic solvent (DES)-emulsification liquid-liquid microextraction of trace curcumin in food and herbal tea samples. *Food Chem.* **2018**, *243*, 442–447. [CrossRef]
22. Ferrone, V.; Genovese, S.; Carlucci, M.; Tiecco, M.; Germani, R.; Preziuso, F.; Epifano, F.; Carlucci, G.; Taddeo, V.A. A green deep eutectic solvent dispersive liquid-liquid micro-extraction (DES-DLLME) for the UHPLC-PDA determination of oxyprenylated phenylpropanoids in olive, soy, peanuts, corn, and sunflower oil. *Food Chem.* **2018**, *245*, 578–585. [CrossRef]
23. Ángeles Fernández, M.D.L.; Espino, M.; Gomez, F.J.; Silva, M.F. Novel approaches mediated by tailor-made green solvents for the extraction of phenolic compounds from agro-food industrial by-products. *Food Chem.* **2018**, *239*, 671–678. [CrossRef] [PubMed]
24. Liew, S.Q.; Ngoh, G.C.; Yusoff, R.; Teoh, W.H. Acid and Deep Eutectic Solvent (DES) extraction of pectin from pomelo (*Citrus grandis* (L.) *Osbeck*) peels. *Biocatal. Agric. Biotechnol.* **2018**, *13*, 1–11. [CrossRef]
25. Hou, X.-D.; Li, A.-L.; Lin, K.-P.; Wang, Y.-Y.; Kuang, Z.-Y.; Cao, S.-L. Insight into the structure-function relationships of deep eutectic solvents during rice straw pretreatment. *Bioresour. Technol.* **2018**, *249*, 261–267. [CrossRef] [PubMed]
26. Jablonský, M.; Škulcová, A.; Kamenská, L.; Vrška, M.; Šíma, J. Deep Eutectic Solvents: Fractionation of Wheat Straw. *BioResources* **2015**, *10*, 8039–8047. [CrossRef]

27. Jablonsky, M.; Haz, A.; Majova, V. Assessing the opportunities for applying deep eutectic solvents for fractionation of beech wood and wheat straw. *Cellulose* **2019**, *26*, 7675–7684. [CrossRef]
28. Jablonský, M.; Majova, V.; Ondrigova, K.; Sima, J. Preparation and characterization of physicochemical properties and application of novel ternary deep eutectic solvents. *Cellulose* **2019**, *26*, 3031–3045. [CrossRef]
29. Škulcova, A.; Haščičová, Z.; Erdlička, L.; Šima, J.; Jablonský, M. Green solvents based on choline chloride for the extraction of spruce bark (*Picea abies*). *Cellulose Chem. Technol.* **2017**, *52*, 3–4.
30. Sakti, A.S.; Saputri, F.C.; Mun'Im, A. Optimization of choline chloride-glycerol based natural deep eutectic solvent for extraction bioactive substances from *Cinnamomum burmannii* barks and *Caesalpinia sappan* heartwoods. *Heliyon* **2019**, *5*, e02915. [CrossRef]
31. Silva, N.H.; Morais, E.; Freire, C.S.; Freire, M.G.; Silvestre, A.J. Extraction of High Value Triterpenic Acids from Eucalyptus globulus Biomass Using Hydrophobic Deep Eutectic Solvents. *Molecules* **2020**, *25*, 210. [CrossRef]
32. Lakka, A.; Grigorakis, S.; Karageorgou, I.; Batra, G.; Kaltsa, O.; Bozinou, E.; Lalas, S.; Makris, D.P. Saffron Processing Wastes as a Bioresource of High-Value Added Compounds: Development of a Green Extraction Process for Polyphenol Recovery Using a Natural Deep Eutectic Solvent. *Antioxidants* **2019**, *8*, 586. [CrossRef]
33. Chen, Z.; Wan, C. Ultrafast fractionation of lignocellulosic biomass by microwave-assisted deep eutectic solvent pretreatment. *Bioresour. Technol.* **2017**, *250*, 532–537. [CrossRef] [PubMed]
34. Alvarez-Vasco, C.; Ma, R.; Quintero, M.; Guo, M.; Geleynse, S.; Ramasamy, K.K.; Wolcott, M.; Zhang, X. Unique low-molecular-weight lignin with high purity extracted from wood by deep eutectic solvents (DES): A source of lignin for valorization. *Green Chem.* **2016**, *18*, 5133–5141. [CrossRef]
35. Skulcova, A.; Jablonsky, M.; Haz, A.; Vrska, M. Pretreatment of wheat straw using deep eutectic solvents and ultrasound. *Przegląd Pap.* **2016**, *72*, 243–247. [CrossRef]
36. Kumar, A.K.; Parikh, B.S.; Pravakar, M. Natural deep eutectic solvent mediated pretreatment of rice straw: Bioanalytical characterization of lignin extract and enzymatic hydrolysis of pretreated biomass residue. *Environ. Sci. Pollut. Res.* **2015**, *23*, 9265–9275. [CrossRef]
37. Sirviö, J.A.; Visanko, M.; Liimatainen, H. Acidic Deep Eutectic Solvents As Hydrolytic Media for Cellulose Nanocrystal Production. *Biomacromolecules* **2016**, *17*, 3025–3032. [CrossRef]
38. Suopajärvi, T.; Sirviö, J.A.; Liimatainen, H. Nanofibrillation of deep eutectic solvent-treated paper and board cellulose pulps. *Carbohydr. Polym.* **2017**, *169*, 167–175. [CrossRef]
39. Xia, Q.; Liu, Y.; Meng, J.; Cheng, W.; Chen, W.; Liu, S.-X.; Liu, Y.; Li, J.; Yu, H. Multiple hydrogen bond coordination in three-constituent deep eutectic solvents enhances lignin fractionation from biomass. *Green Chem.* **2018**, *20*, 2711–2721. [CrossRef]
40. Xu, G.; Ding, J.-C.; Han, R.-Z.; Dong, J.-J.; Ni, Y. Enhancing cellulose accessibility of corn stover by deep eutectic solvent pretreatment for butanol fermentation. *Bioresour Technol.* **2016**, *203*, 364–369. [CrossRef]
41. Ruesgas-Ramón, M.; Suárez-Quiroz, M.L.; González-Ríos, O.; Baréa, B.; Cazals, G.; Figueroa-Espinoza, M.C.; Durand, E. Biomolecules extraction from coffee and cocoa by- and co-products using deep eutectic solvents. *J. Sci. Food Agric.* **2019**, *100*, 81–91. [CrossRef]
42. Ljekocevic, M.; Jadranin, M.; Stankovic, J.; Popovic, B.; Nikicevic, N.; Petrovic, A.; Tešević, V. Phenolic composition and anti-DPPH radical scavenging activity of plum wine produced from three plum cultivars. *J. Serbian Chem. Soc* **2019**, *84*, 141–151. [CrossRef]
43. Hidalgo, G.I.; Almajano, M.P. Red Fruits: Extraction of Antioxidants, Phenolic Content, and Radical Scavenging Determination: A Review. *Antioxidants* **2017**, *6*, 7. [CrossRef] [PubMed]
44. Yunusa, A.K.; Rohin, M.A.K.; Bakar, C.H.A.B. Free radical scavenging activity of polyphenols. *J. Chem. Pharm. Res.* **2015**, *7*, 1975–1980.
45. Lu, Y.; Foo, L.Y. Antioxidant and radical scavenging activities of polyphenols from apple pomace. *Food Chem.* **2000**, *68*, 81–85. [CrossRef]
46. Anwar, H.; Hussain, G.; Mustafa, I. Antioxidants from Natural Sources. In *Antioxidants in Foods and Its Applications*; IntechOpen: Rijeka, Croatia, 2018; pp. 1–27.
47. Alcalde, B.; Granados, M.; Saurina, J. Exploring the Antioxidant Features of Polyphenols by Spectroscopic and Electrochemical Methods. *Antioxidants* **2019**, *8*, 523. [CrossRef]
48. Makris, D.P.; Şahin, S. Polyphenolic Antioxidants from Agri-Food Waste Biomass. *Antioxidants* **2019**, *8*, 624. [CrossRef]

49. Silva, R.F.M.; Pogačnik, L. Polyphenols from Food and Natural Products: Neuroprotection and Safety. *Antioxidants* **2020**, *9*, 61. [CrossRef]
50. Selvasundhari, L.; Babu, V.; Jenifer, V.; Jeyasudha, S.; Thiruneelakandan, G.; Sivakami, R.; Anthoni, S. In Vitro Antioxidant Activity of Bark Extracts of *Rhizophora mucronata*. *Sci. Technol. Arts Res. J.* **2014**, *3*, 21. [CrossRef]
51. Patrick, A.T.; Samson, F.P.; Jalo, K.; Thagriki, D.; Umaru, H.A.; Madusolumuo, M.A. In vitro antioxidant activity and phytochemical evaluation of aqueous and methanolic stem bark extracts of *Pterocarpus erinaceus*. *World J. Pharm. Res.* **2016**, *5*, 134–151.
52. Das, S.; Ray, A.; Nasim, N.; Nayak, S.; Mohanty, S. Effect of different extraction techniques on total phenolic and flavonoid contents, and antioxidant activity of betelvine and quantification of its phenolic constituents by validated HPTLC method. *3 Biotech* **2019**, *9*, 37. [CrossRef]
53. Strižincová, P.;Ház, A.; Burčová, Z.; Feranc, J.; Kreps, F.; Šurina, I.; Jablonský, M. Spruce Bark-A Source of Polyphenolic Compounds: Optimizing the Operating Conditions of Supercritical Carbon Dioxide Extraction. *Molecules* **2019**, *24*, 4049. [CrossRef]
54. Burcova, Z.; Kreps, F.; Strizincova, P.; Haz, A.; Jablonsky, M.; Surina, I.; Schmidt, S. Spruce Bark as a Source of Antioxidant Active Substances. *BioResources* **2019**, *14*, 5980–5987.
55. Kreps, F.; Burčová, Z.; Jablonský, M.;Ház, A.; Frecer, V.; Kyselka, J.; Schmidt, Š.; Šurina, I.; Filip, V. Bioresource of Antioxidant and Potential Medicinal Compounds from Waste Biomass of Spruce. *ACS Sustain. Chem. Eng.* **2017**, *5*, 8161–8170. [CrossRef]
56. Ferreira-Santos, P.; Genisheva, Z.; Pereira, R.N.; Teixeira, J.A.; Rocha, C.M. Moderate Electric Fields as a Potential Tool for Sustainable Recovery of Phenolic Compounds from Pinus pinaster Bark. *ACS Sustain. Chem. Eng.* **2019**, *7*, 8816–8826. [CrossRef]
57. Lazar, L.; Talmaciu, A.I.; Volf, I.; Popa, V.I. Kinetic modeling of the ultrasound-assisted extraction of polyphenols from Picea abies bark. *Ultrason. Sonochem.* **2016**, *32*, 191–197. [CrossRef] [PubMed]
58. Conde, E.; Hemming, J.; Smeds, A.; Reinoso, B.D.; Moure, A.; Willför, S.; Domínguez, H.; Parajó, J. Extraction of low-molar-mass phenolics and lipophilic compounds from Pinus pinaster wood with compressed CO_2. *J. Supercrit. Fluids* **2013**, *81*, 193–199. [CrossRef]
59. Chupin, L.; Maunu, S.L.; Reynaud, S.; Pizzi, A.; Charrier, B.; Bouhtoury, F.C.-E. Microwave assisted extraction of maritime pine (*Pinus pinaster*) bark: Impact of particle size and characterization. *Ind. Crop. Prod.* **2015**, *65*, 142–149. [CrossRef]
60. Vieito, C.; Fernandes, É.; Velho, M.V.; Pires, P. The effect of different solvents on extraction yield, total phenolic content and antioxidant activity of extracts from pine bark (*Pinus pinaster* subsp. atlantica). *Chem. Eng. Trans.* **2018**, *64*, 127–132.
61. Jablonsky, M.; Haz, A.; Burcova, Z.; Kreps, F.; Jablonsky, J. Pharmacokinetic properties of biomass-extracted substances isolated by green solvents. *BioResources* **2019**, *14*, 6294–6303.
62. Haz, A.; Strizincova, P.; Majova, V.; Skulcova, A.; Jablonsky, M. Thermal stability of selected deep eutectic solvents. *Int. J. Recent. Sci. Res.* **2016**, *7*, 14441–14444.
63. Lynam, J.; Kumar, N.; Wong, M.J. Deep eutectic solvents' ability to solubilize lignin, cellulose, and hemicellulose; thermal stability; and density. *Bioresour. Technol.* **2017**, *238*, 684–689. [CrossRef]
64. Hromadkova, Z.; Paulsen, B.S.; Polovka, M.; Kostalova, Z.; Ebringerova, A. Structural features of two heteroxylan polysaccharide fractions from wheat bran with anti-complementary and antioxidant activities. *Carbohydr. Polym.* **2013**, *93*, 22–30. [CrossRef] [PubMed]
65. Martins, M.A.R.; Pinho, S.P.; Coutinho, J.A.P. Insights into the Nature of Eutectic and Deep Eutectic Mixtures. *J. Solut. Chem.* **2018**, *48*, 962–982. [CrossRef]
66. Francisco, M.; Bruinhorst, A.V.D.; Kroon, M.C. Low-Transition-Temperature Mixtures (LTTMs): A New Generation of Designer Solvents. *Angew. Chem. Int. Ed.* **2013**, *52*, 3074–3085. [CrossRef] [PubMed]
67. Smith, P.; Arroyo, C.B.; Hernandez, F.L.; Goeltz, J.C. Ternary Deep Eutectic Solvent Behavior of Water and Urea? Choline Chloride Mixtures. *J. Phys. Chem. B* **2019**, *123*, 5302–5306. [CrossRef] [PubMed]
68. Kähkönen, M.P.; Hopia, A.I.; Vuorela, H.; Rauha, J.-P.; Pihlaja, K.; Kujala, T.S.; Heinonen, M. Antioxidant Activity of Plant Extracts Containing Phenolic Compounds. *J. Agric. Food Chem.* **1999**, *47*, 3954–3962. [CrossRef]
69. Karppanen, O.; Venäläinen, M.; Harju, A.M.; Laakso, T. The effect of brown-rot decay on water adsorption and chemical composition of Scots pine heartwood. *Ann. For. Sci.* **2008**, *65*, 610. [CrossRef]

70. Siren, H.; Kaijanen, L.; Kaartinen, S.; Väre, M.; Riikonen, P.; Jernström, E. Determination of statins by gas chromatography—EI/MRM—Tandem mass spectrometry: Fermentation of pine samples with Pleurotus ostreatus. *J. Pharm. Biomed. Anal.* **2014**, *94*, 196–202. [CrossRef]
71. Willför, S.; Hemming, J.; Reunanen, M.; Holmbom, B. Phenolic and Lipophilic Extractives in Scots Pine Knots and Stemwood. *Holzforschung* **2003**, *57*, 359–372. [CrossRef]
72. Zhao, Z.; Moghadasian, M. Chemistry, natural sources, dietary intake and pharmacokinetic properties of ferulic acid: A review. *Food Chem.* **2008**, *109*, 691–702. [CrossRef]
73. Agra, L.C.; Ferro, J.N.S.; Barbosa, F.T.; Barreto, E. Triterpenes with healing activity: A systematic review. *J. Dermatol. Treat.* **2015**, *26*, 1–6. [CrossRef]
74. Talmaciu, A.I.; Volf, I.; Popa, V.I. Supercritical fluids and ultrasoud assisted extractions applied to spruce bark conversion. *Environ. Eng. Manag. J. (EEMJ)* **2015**, *14*, 615–623.
75. Tripoli, E.; La Guardia, M.; Giammanco, S.; Di Majo, D.; Giammanco, M. Citrus flavonoids: Molecular structure, biological activity and nutritional properties: A review. *Food Chem.* **2007**, *104*, 466–479. [CrossRef]
76. Coşarcă, S.-L.; Moacă, E.-A.; Tanase, C.; Muntean, D.L.; Pavel, I.Z.; Dehelean, C.A. Spruce and beech bark aqueous extracts: Source of polyphenols, tannins and antioxidants correlated to in vitro antitumor potential on two different cell lines. *Wood Sci. Technol.* **2018**, *53*, 313–333. [CrossRef]
77. Sládková, A.; Benedeková, M.; Stopka, J.; Šurina, I.; Ház, A.; Strižincová, P.; Čižová, K.; Škulcová, A.; Burčová, Z.; Kreps, F.; et al. Yield of Polyphenolic Substances Extracted from Spruce (*Picea abies*) Bark by Microwave-Assisted Extraction. *BioResources* **2016**, *11*, 9912–9921. [CrossRef]
78. Huang, Y.; Feng, F.; Jiang, J.; Qiao, Y.; Wu, T.; Voglmeir, J.; Chen, Z.-G. Green and efficient extraction of rutin from tartary buckwheat hull by using natural deep eutectic solvents. *Food Chem.* **2017**, *221*, 1400–1405. [CrossRef]
79. Radošević, K.; Ćurko, N.; Srček, V.G.; Bubalo, M.C.; Tomasevic, M.; Ganić, K.K.; Redovniković, I.R. Natural deep eutectic solvents as beneficial extractants for enhancement of plant extracts bioactivity. *LWT* **2016**, *73*, 45–51. [CrossRef]
80. Nam, M.W.; Lee, J.; Zhao, J.; Jeong, J.H. Enhanced extraction of bioactive natural products using tailor-made deep eutectic solvents: Application to flavonoid extraction from Flos sophorae. *Green Chem.* **2015**, *17*, 1718–1727. [CrossRef]
81. Bakirtzi, C.; Triantafyllidou, K.; Makris, D.P. Novel lactic acid-based natural deep eutectic solvents: Efficiency in the ultrasound-assisted extraction of antioxidant polyphenols from common native Greek medicinal plants. *J. Appl. Res. Med. Aromat. Plants* **2016**, *3*, 120–127. [CrossRef]
82. Predescu, N.C.; Papuc, C.; Nicorescu, V.; Gajaila, I.; Goran, G.V.; Petcu, C.D.; Stefan, G. The influence of solid-to-solvent ratio and extraction method on total phenolic content, flavonoid content and antioxidant properties of some ethanolic plant extracts. *Rev. Chim.* **2016**, *67*, 1922–1927.
83. Mahattanatawee, K.; Manthey, J.A.; Luzio, G.; Talcott, S.T.; Goodner, K.; Baldwin, E.A. Total Antioxidant Activity and Fiber Content of Select Florida-Grown Tropical Fruits. *J. Agric. Food Chem.* **2006**, *54*, 7355–7363. [CrossRef]
84. Díaz-Maroto, M.C.; Palomo, L.; Rodríguez, L.; Fuentes-Contreras, E.; Arráez-Román, D.; Segura-Carretero, A. Antiplatelet Activity of Natural Bioactive Extracts from Mango (*Mangifera Indica* L.) and its By-Products. *Antioxidants* **2019**, *8*, 517. [CrossRef]
85. Di Mauro, M.D.; Fava, G.; Spampinato, M.; Aleo, D.; Melilli, B.; Saita, M.G.; Centonze, G.; Maggiore, R.; D'Antona, N. Polyphenolic Fraction from Olive Mill Wastewater: Scale-Up and in Vitro Studies for Ophthalmic Nutraceutical Applications. *Antioxidants* **2019**, *8*, 462. [CrossRef] [PubMed]
86. Fernandes, P.; Ferreira, S.; Bastos, R.; Ferreira, I.; Cruz, M.; Pinto, A.; Coelho, E.; Passos, C.; Coimbra, M.; Cardoso, S.; et al. Apple pomace extract as a sustainable food ingredient. *Antioxidants* **2019**, *8*, 189. [CrossRef]
87. Li, Y.; Bundeesomchok, K.; Rakotomanomana, N.; Fabiano-Tixier, A.-S.; Bott, R.; Wang, Y.; Chemat, F. Towards a Zero-Waste Biorefinery Using Edible Oils as Solvents for the Green Extraction of Volatile and Non-Volatile Bioactive Compounds from Rosemary. *Antioxidants* **2019**, *8*, 140. [CrossRef]
88. Buchman, A.L. The Addition of Choline to Parenteral Nutrition. *Gastroenterology* **2009**, *137*, S119–S123. [CrossRef] [PubMed]

89. Zhuang, B.; Dou, L.-L.; Li, P.; Liu, E.-H. Deep eutectic solvents as green media for extraction of flavonoid glycosides and aglycones from *Platycladi Cacumen*. *J. Pharm. Biomed. Anal.* **2017**, *134*, 214–219. [CrossRef]
90. Karageorgou, I.; Grigorakis, S.; Lalas, S.; Makris, D.P. Enhanced extraction of antioxidant polyphenols from Moringa oleifera Lam. leaves using a biomolecule-based low-transition temperature mixture. *Eur. Food Res. Technol.* **2017**, *21*, 17–1848. [CrossRef]
91. Li, D.P.; Calvert, P.; Mello, C.; Morabito, K.; Tripathi, A.; Shapley, N.; Gilida, K. Stabilization of Natural Dyes by High Levels of Antioxidants. *Adv. Mater. Res.* **2012**, *441*, 192–199. [CrossRef]
92. Li, Y.-D.; Guan, J.-P.; Tang, R.-C.; Qiao, Y.-F. Application of Natural Flavonoids to Impart Antioxidant and Antibacterial Activities to Polyamide Fiber for Health Care Applications. *Antioxidants* **2019**, *8*, 301. [CrossRef]
93. Varatharajan, K.; Rani, D. Screening of antioxidant additives for biodiesel fuels. *Renew. Sustain. Energy Rev.* **2018**, *82*, 2017–2028. [CrossRef]
94. García, M.; Botella, L.; Gil-Lalaguna, N.; Arauzo, J.; Gonzalo, A.; Sánchez, J. Antioxidants for biodiesel: Additives prepared from extracted fractions of bio-oil. *Fuel Process. Technol.* **2017**, *156*, 407–414. [CrossRef]
95. Fiorio, R.; D'Hooge, D.R.; Ragaert, K.; Cardon, L. A Statistical Analysis on the Effect of Antioxidants on the Thermal-Oxidative Stability of Commercial Mass- and Emulsion-Polymerized ABS. *Polymers* **2018**, *11*, 25. [CrossRef] [PubMed]
96. Boersma, A. Predicting the efficiency of antioxidants in polymers. *Polym. Degrad. Stab.* **2006**, *91*, 472–478. [CrossRef]
97. Abdalla, H.S.; Patel, S. The performance and oxidation stability of sustainable metalworking fluid derived from vegetable extracts. *Proc. Inst. Mech. Eng. Part B J. Eng. Manuf.* **2006**, *220*, 2027–2040. [CrossRef]
98. Nimse, S.B.; Pal, D. Free radicals, natural antioxidants, and their reaction mechanisms. *RSC Adv.* **2015**, *5*, 27986–28006. [CrossRef]

© 2020 by the authors. Licensee MDPI, Basel, Switzerland. This article is an open access article distributed under the terms and conditions of the Creative Commons Attribution (CC BY) license (http://creativecommons.org/licenses/by/4.0/).

Review

Phytomass Valorization by Deep Eutectic Solvents—Achievements, Perspectives, and Limitations

Michal Jablonský [1,*] and Jozef Šima [2]

[1] Department of Chemical Technology of Wood, Pulp and Paper, Faculty of Chemical and Food Technology, Slovak University of Technology in Bratislava, Radlinského 9, SK-812 37 Bratislava, Slovakia
[2] Department of Inorganic Chemistry, Faculty of Chemical and Food Technology, Slovak University of Technology in Bratislava, Radlinského 9, SK-812 37 Bratislava, Slovakia; jozef.sima@stuba.sk
* Correspondence: michal.jablonsky@stuba.sk

Received: 7 August 2020; Accepted: 9 September 2020; Published: 10 September 2020

Abstract: In recent years, a plethora of extraction processes have been performed by a novel class of green solvents known as deep eutectic solvents (DESs), possessing several environmental, operational, and economic advantages proven by experience when compared to organic solvents and ionic liquids. The present review provides an organized overview of the use of DESs as extraction agents for the recovery of valuable substances and compounds from the original plant biomass, waste from its processing, and waste from the production and consumption of plant-based food. For the sake of simplicity and speed of orientation, the data are, as far as possible, arranged in a table in alphabetical order of the extracted substances. However, in some cases, the isolation of several substances is described in one paper and they are, therefore, listed together. The table further contains a description of the extracted phytomass, DES composition, extraction conditions, and literature sources. With regard to extracted value-added substances, this review addresses their pharmacological, therapeutic, and nutritional aspects. The review also includes an evaluation of the possibilities and limitations of using DESs to obtain value-added substances from phytomass.

Keywords: deep eutectic solvents; phytomass; valorization; extraction

1. Introduction

Biomass is considered to be any organic material produced by the growth of microorganisms, plants, or animals, involving also wastes and residues of organic nature [1,2]. Biomass is a renewable energy source, the second oldest source of energy following the Sun. Primary biomass and biowaste generated during its treatment, processing, and use are the source of a huge number of compounds and substances, referred to as value-added products, which can be extracted, recovered, and/or synthesized from biomass [3]. The ways of obtaining such value-added products are covered under the term valorization. The sum of the biomass across all taxa on Earth is approximately 550 gigatons of carbon, of which about 450 Gt C are plants, followed by bacteria (70 Gt C), fungi, archaea, protists, animals, and viruses, which together account for the remaining <10% [4].

Based on the composition of biomass, it is understandable that the attention of researchers and technologists is focused on plants and plant waste, which we will refer to as phytomass. The main mode of obtaining value-added products from phytomass and discussed in this review are extraction processes. In accordance with ecologically oriented developments in current chemistry and technology, we will deal with green extraction agents, specifically deep eutectic solvents.

This work is aimed at providing an overview of the obtained value-added products, their phytomass sources, used DESs, and conditions of valorization. The work is mainly devoted to results achieved in the last few years and indicates the perspectives, but also the limitations of the development of this area.

2. Deep Eutectic Solvents

Over time, several types of extractants have been used to recover compounds and substances from phytomass. At the beginning, it was water, later on followed by organic solvents. At the end of the previous millennium, efforts to protect the environment and obtain cleaner products in higher yields in a less costly manner led to new kinds of extractants—ionic liquids. Probably the first room-temperature ionic liquid was described as early as 1914 [5]. Ionic liquids were utilized also in valorization of phytomass, specifically in obtaining some compounds from cellulose [6,7].

Some unfavorable characteristics of ionic liquids (sensitivity to humidity; toxicity; production, handling, and disposal costs; non-biodegradability; inflammability) have been overcome by a newer type of solvent—low-transition temperature mixtures (LTTMs). From the view of applicability, cost-relating factor is of key importance. The cost of producing a DES has been estimated to be 20% of that of an ionic liquid [8]. LTTMs are mixtures of two or more high-melting-point starting materials—hydrogen bond donor (HBD) and hydrogen bond acceptor (HB)—which can bond with each other to form a mixture having a final melting point that is lower than that of the starting components, becoming thus, usually liquids at room temperature. LTTMs are composed of inexpensive, recyclable, and non-toxic materials, frequently of natural origin (e.g., sugars, organic acids, and salts, etc.). In the field of these green solvents and extractants, along with the term LTTMs, various terms have been introduced, such as deep eutectic solvents (DESs), natural deep eutectic solvents (NADESs), and low-melting mixtures (LMMs). As pointed out by several scientific and industrial teams [9–15], it is obvious that these terms are interchangeable and there is no substantial difference in them. In this paper, we will stick to the term DESs coined by Abbott et al. in their pioneering paper [16] for liquids composed of natural high-melting-point starting materials.

It should be emphasized that despite the seemingly similar extraction effects of ionic liquids and DESs, there is a fundamental chemical difference between these types of extractants. Ionic liquids are ionic compounds; their components are ions attracted by the ionic bond. Components of DESs are bound by the hydrogen bond and van der Waals interactions; DESs are thus just mixtures. The fact that the composition of HBD and HBA is retained in the liquid phase after mixing allows for easy regeneration of DESs after use as a solvent.

Although many DESs are not eutectic mixtures in the exact sense of the word, the use of the acronym DES is constantly expanding, probably due to the fact that they are liquids at room temperature, they can be easily prepared without the need for complex and expensive cleaning procedures, are used in an environmentally friendly manner, and easily regenerate [15].

Based on Abbott's fundamental works and reflecting the nature of starting components, DESs are classified into five types [17,18]: Type I (quaternary salt and metal halide), Type II (quaternary salt and hydrated metal halide), Type III (quaternary salt and hydrogen bond donor), Type IV (metal halide and hydrogen bond donor), and Type V (non-ionic DESs composed only of molecular substances). Type V DESs were defined by Abranches et al. in 2019 [18], although the first non-ionic deep eutectic mixture was described by Usanovich as early as 1958 [19]. Such non-ionic DESs have found their application in polymer chemistry, medicine [20], and other application areas [21,22]. They have not been systematically used as extractants so far. When handling materials of biological origin, it is advantageous to avoid metal-containing compounds, and therefore, DESs of type III (and probably, Type V in the future) are preferred in the valorization of phytomass.

The basic properties of two- or more component DESs have been described by several authors [9,17,23–30]. The knowledge gained so far can be summarized as follows. DESs are biodegradable, cheap, easy to prepare, low toxic, fire resistant, miscible with water, have negligible vapor pressure, and are liquid in a wide temperature range. For most Type III DESs, their room-temperature viscosity (19–13,000 cP) and density (1.0 to 1.4 g/cm^3) are higher than that of water and their electrical conductivity is rather low (<1 mS/cm). All of the mentioned physicochemical characteristics are temperature-dependent. Until recently, only hydrophilic DESs composed of typically hydrophilic materials were available [8–11]. The first data on hydrophobic DESs, stable in contact with water, were published in 2015 [10] and reviewed in 2019 [31].

Being multicomponents systems, DESs offer significant advantages over conventional solvents: their structure may be modified by the selection of solvent-forming components, as well as by the molar ratio of the components participating in hydrogen bond formation. That is why their properties (e.g., melting/glass transition temperature, viscosity, conductivity, refractive index, density, and pH) are significantly influenced by the DESs composition and can be purposefully modified to some extent. Given the high purity of their components, DESs can be prepared in the form of high purity mixtures in waste-free processes. One of the most attractive properties of DESs is their biodegradability, based on the use of natural precursors. All these characteristics caused DESs to be proposed instead of common organic volatile solvents, preventing, thus, a large-scale release of flammable vapors and emissions. In addition, the release of solvent during reaction processes, separation, and purification becomes limited.

3. Deep Eutectic Solvents as Extracting Agents

3.1. Requirements for Extracting Agents

DESs have been introduced as extractants in several areas. Their ability to function as denitrification agents was documented by Rogošic et al. [32]. Adjusting the HBA:HBD molar ratio, several DESs documented their advantage in analysis and separation of organic and inorganic compounds from food samples [33]. DESs composed of tetrabutylammonium bromide and long-chain alcohols were investigated as extracting solvents for headspace single-drop microextraction of more than 40 terpenes from six spices (cinnamon, cumin, fennel, clove, thyme, and nutmeg). Advantages in extraction of metals from mixed oxide matrixes were described by the team headed by Abbott [34]. Triaux and his coworkers [35] found that the most important factors for separation efficiency were extraction time and temperature. Along with their application in chromatography and biomass processing, some illustrative results of extraction of value-added compounds from biomass are described too [8,36–38].

The advantages of DESs as extractants over organic solvents and ionic liquids are mentioned in a nutshell above. To fulfil the main requirements of valorization, i.e., to obtain the highest possible amount of the desired compounds in the highest purity at the lowest total cost and adverse environmental impact, both the treated biomass and the extractant must meet optimal parameters. Among the properties of DESs, their thermal stability, liquid state temperature range, viscosity, polarity, and acidic basic properties are particularly important in the search for optimal DESs for the extraction of selected compounds. Other traditionally measured and published properties, such as electrical conductivity, refractive index, and density, are not decisive in selecting DESs for separation and extraction purposes.

3.2. Thermal Stability

Thermal stability is a key requirement of DESs in assessing the suitability of their use as extractants [11]. Thermal stability is defined either at the molecular level as the stability of a molecule when it is exposed to very high temperature or at the substance level as the resistance of a compound to decomposition and/or loss of mass at high temperatures [28]. From the viewpoint of practical application, thermal stability of a DES means a measure of how long the DES can hold before dissociating into its components and/or the breaking down by heat into smaller decomposition products which do not recombine on cooling. Thermal stability as part of thermal properties is usually monitored by thermogravimetric analysis. This technique allows researchers and technologists to determine temperature intervals of structural compositional changes of substances and weight loss. As far as DESs are concerned, their decomposition temperature is the highest temperature of their practical use. It should be noted at this point that increasing the extraction temperature is limited not only by the thermal stability of the used DES but also by the processed biomass and extracted compounds.

3.3. Temperature Range of Liquid State

When evaluating possibilities of the use of organic solvents as extractants, the temperature range of their liquid state is one of the key parameters. However, this is not the case with DESs. Even in otherwise excellent papers and reviews devoted to DESs [9,11,17,23–30,39,40], instead of the temperature range of liquid state, only the melting/glass transition temperature is given. The reason for this situation is understandable since the boiling point of DESs usually is not measurable due to their decomposition at higher temperatures.

It should be pointed out also that published melting/glass transition temperatures must be taken with some caution. From the theoretical point of view, the significant melting point depression of a eutectic mixture compared to that of the pure HBA and HBD is due to several factors, such as charge delocalization (from, for instance, the halide anion (HBA) to the HBD, facilitated by hydrogen bond formation); a reduction in strength of several other cohesive interactions counterbalancing the increased strength of the H-bonds developed at a eutectic composition; the lattice energy of the HBA and HBD; the way the anion and HBD interact; and the entropy changes upon DES formation. From the practical point of view, problems lie in the fact that the mentioned temperatures depend to a large extent on the purity of the DES and the water content. The determined properties of DESs prepared from pure water-free components on a laboratory scale may not be identical to the properties of large-scale applied DESs.

3.4. Viscosity

To extract a desired compound from phytomass, there must be a contact of an extractant (DES, in our case) and an extractable compound. The penetration of DES into the body of phytomass strongly depends on DES viscosity. Generally speaking, the higher its extractant viscosity, the lower its extractive efficiency. Viscosity can be decreased in three main ways: by changing the molar ratio of the DES components (HBA and HBD); by adding water, organic solvent, or another additive; by rising the temperature. Viscosity data can be found in [11,17,23,26,41]. For several DESs, viscosity/temperature dependences are expressed by mathematical equations [11]; viscosity/DES composition dependences are individual for each DES. More details on viscosity issues are given in part "Factors limiting potential of deep eutectic solvents utilization and how to overcome them".

3.5. Polarity

The importance and effect of polarity in the separation of individual components from a multicomponent system are well-known in chromatographic methods. To express the polarity of chemical substances and their mixtures including DESs, several parameters are used. The classic, most frequently used quantities are relative permittivity εr (dielectric constant) [42], spectral parameter $E_T(30)$, and Kamlet–Taft π^*, α, and β parameters. For pure organic solvents, several polarity parameters based on equilibrium, kinetic, and spectroscopic measurements are discussed in detail by Reichardt and Welton [43]. It should be pointed out that polarity is not an absolute property of the pure liquid and hence, there is no single correct value when comparing polarity scales. All polarity scales are relative and different scales give different polarities for the same solvent [44–46].

As far as the polarity of DESs is concerned, the literature is very sparse in data. Even in an excellent book on DESs [47], only a few lines are devoted to the polarity of DESs. Given the fact that DESs are composed of HBA and HBD, it can be expected that as the most suitable parameters characterizing DESs, Kamlet–Taft parameters π^*, α, and β will be used preferentially. The parameters α and β express the H-bond acidity of HBD and basicity of HBA, respectively; π^* is a measure of dipolarity/polarizability of the solvents.

Measurements of choline chloride (ChCl)-based DESs with urea, glycerol, acetic acid, malonic acid, glycolic acid, ethylene glycol, or levulinic acid, as well as those of DESs composed of tetrabutylammonium chloride and levulinic acid, indicate that their polarity is close to that of water [46,48]. The fact that the composition of two- or multicomponent DESs can be varied almost indefinitely in practice opens up the chance to prepare a DES with the required polarity suitable for the desired applications. The ability to modify polarity can be expected to be very important in the separation of lipophilic (non-polar) and hydrophilic (polar) nutrients from phytomass [37].

Taking the polarity aspect into account, it is worth pointing out the ability of some solvents to exhibit switchable polarity [49]. Switchable polarity solvents equilibrate between a higher polarity and a lower polarity when a trigger is applied. These solvents are particularly useful when two different polarities of the solvent are needed for two different steps. Up to now, mainly ionic liquids have been considered as belonging to the category of switchable polarity solvents; there is, however, no fundamental reason hampering the introduction of DESs into the same category. Despite the assumption of the importance of the polarity of DESs for their extraction properties, this phenomenon has not yet received the necessary attention.

3.6. Acid-Base Properties

The extractability of compounds from any raw materials, including phytomass, may depend, along with the nature of extracted compounds, to various extents on the acid-base properties of the DES used. In general, the acidity or basicity of DESs is determined mainly by the acidity/basicity constants and molar ratio of their components. It is, therefore, more or less predictable. Values of pH for the aqueous solutions of selected DESs range from basic DES with pH ≈ 13 (a DES containing glycerol and K_2CO_3, [50]), through nearly neutral DESs, to acidic ones containing an organic acid [28].

The predominance of acid-catalyzed reactions in synthetic chemistry has led researchers to focus more on acidic ionic liquids and DESs, as well as on unveiling modes of tailoring their properties. Two categories of acidic DESs were formulated, namely Brønsted acidic DESs displaying Brønsted acidity due to ionizable protons, and Lewis acidic DESs displaying Lewis acidity because of a deficiency in electron [39]. This categorization has not yet been introduced into the field of phytomass valorization by DESs.

As pointed out by Trajano and Wyman [51], along with advantages of some reactions performed at a lower pH (such as higher yields of desired compounds), there are also drawbacks concerning the necessity to use corrosion-proof equipment. Acidic DESs are able to react with some metals and dissolve their oxides. Moreover, the products must frequently be washed and neutralized.

4. Therapeutic Effects of Substances Extracted from Phytomass

Since the beginning of civilization, man has been associated with plants and herbs and has used their potential in the treatment of various diseases. Without knowing the plants' components, he learned what types are suitable for which diseases and how to prepare adequate preparations from these plants, which are used to protect or restore health, to alleviate disease manifestations but also to recognize the disease. The therapeutic potential of substances contained in medicinal plants has been historically proven [52]. The world's population (80%) is engaged in folk medicine based on the use of plants [53]. Secondary metabolites are the most successful source of potential drugs. Herbal-derived chemicals are known to decrease the risk of some severe disorders, including autoimmune and cardiovascular diseases, as well as neurodegenerative diseases [54].

Many plants contain a wide range of inhibitors of viral proteins and act against viral diseases. Plants can generally produce metabolites that have an inhibitory effect on the proliferation of enzymes, proteins, and viruses [55,56]. Natural products have the potential to form the basis of holistic healthcare. For some people, synthetic drugs cause harmful side effects and are expensive to buy compared to traditional herbal products [57], although "natural" does not automatically mean "harmlessness".

The therapeutic potential of herbal medicines was assessed in a variety of animal models, and their effect and mechanisms of action were investigated through neurochemical approaches [58].

There is scientific evidence and centuries of human empirical experience on the therapeutic superiority of plant extracts over individual isolated ingredients, as well as their biological equivalence with synthetic drugs. The results of various studies on the effects of multicomponent extracts are summarized in the Wagner study [59]. In recent decades, pharmaceutical research and industry have sought to uncover the causes of the pharmacological and therapeutic superiority of many natural multi-ingredient products over individual compounds. One of the explanations is that plant extracts may contain bioactive natural products in the form of prodrugs, and in some cases, these compounds may optimize the effects of individual substances to achieve therapeutic goals. An illustrative example was provided by David et al. [57] documenting that in plants, many natural products exist in the form of conjugates with saccharide moieties (called glycosides). Many of the glycosides are activated upon cell disruption to yield highly active therapeutic compounds (e.g., glucosinolates and cyanogenetic glycosides). Thus, the glycosides themselves are not active directly; however, they can become active and efficient upon metabolization.

Another explanation lies in the existence of a synergistic therapeutic effect of several active substances in natural extracts. Flavonoids have a wide variety of biological activities and therapeutic potential [60]. Ginkgolide A and B can serve as examples. It is known that the combination of ginkgolide A and B acts on antiplatelet-activating factors in ginkgo biloba phytopharmaceuticals. The synergistic effect of the combination of these ginkgolides in the preparation of ginkgo biloba was confirmed by Wagner [61]. The extracts may contain several bioactive compounds with different specific activity. Polyphenols extracted by DESs such as curcumin, ferulic acid, proanthocyanidin, quercetin, quercetin-3-O-glucoside, isorhamnetin, kaempferol, rutin, p-coumaric acid, gallic acid, caffeic acid, catechin, epicatechin, catechinic acid, chlorogenic acid, syringic acid, vanillic acid, and others have shown antioxidant, anti-inflammatory properties [62]. Aromatic phytochemical constituents have analgesic, anticarcinogenic, antidiabetic, antifungal, cardioprotective, hepatoprotective, and neuroprotective properties [52,54,62]. The potential of the 237 extracted substances from phytomass for orally bioavailable therapeutics by predicting a number of ADME (Absorption, Distribution, Metabolism, and Extraction)-related properties was published in the work Jablonsky et al. 2017 [62]. When using such multicomponent extracts, several therapeutic effects of their individual components may be exerted. In order to isolate the desired secondary plant metabolites, it is possible to utilize various extraction techniques. Based on the chosen extraction method, it is subsequently possible to yield various types of substances in various quality and composition.

The interest in natural products or obtaining active substances from plants also has an economic background. This is based on society's efforts and beliefs about the benefits of returning to traditional medicine. This had the effect that herbal phytopharmaceuticals reached USD 60 billion, with annual growth rates of 5–15% [63].

5. Valorization of Phytomass by Deep Eutectic Solvents

Extraction Techniques

To date, a number of papers have been published on the extraction and separation of value-added compounds and substances from phytomass. In order to achieve the maximum yield, purity, and selectivity of such substances, their extraction from phytomass is carried out by purposefully selected methods. Their choice depends predominantly on the processed phytomass and required target compounds. Along with classical auxiliary techniques (heating, centrifugation, shaking), the application of DESs becomes associated with advanced extraction techniques, such as ultrasound-assisted extraction (UAE), negative pressure cavitation (NPC), enzyme-assisted extraction (EAE), hydrodiffusion extraction (HDE), supercritical fluid extraction (SFE), and microwave-assisted extraction (MAE). The mentioned techniques are thoroughly evaluated in the review published by Cunha and Fernandes [64]. The extraction of value-added compounds is often associated with the pretreatment of phytomass, which means the treatment of the inputted raw phytomass, e.g., by mechanical milling prior to the action of DES as extractants to facilitate the penetration of DES into the processed phytomass and to improve the contact of DES with the extracted components by disrupting the solid impermeable structures of the natural polymers [8].

6. Extracted Value-Added Compounds

The achievements in obtaining value-added compounds by phytomass valorization with a focus on the last 5 years are shown in Table 1. For clarity and convenient orientation, the data are arranged, as far as possible, in alphabetical order of the extracted compounds.

Table 1. Value-added compounds extracted from phytomass by Type III DESs and extraction conditions.

Compounds	Sample	Solvent	Ratio	Extraction Conditions; Analytical Methods	Ref.
Acacetin-7-diglucuronide, Apigenin-7-O-diglucuronide, Campneoside, Cistanoside F, Dimethyl quercetin, Durantoside I, Eukovoside, Forsythoside A, Gardoside, Chrysoeriol-7-diglucuronide, Isoverbascoside, Ixoside, Lippioside, Lippioside I derivative, Lippioside II, Luteolin-7-diglucuronide, Martynoside, Oxoverbascoside, Teucardoside, Theveside, Verbascoside, β-hydroxyverbascoside/β-hydroxy-iso verbascoside, Total phenolic compounds	*Lippia citriodora*	ChCl:lactic acid ChCl:tartaric acid ChCl:1,3-batanediol ChCl:ethylene glycol ChCl:xylitol ChCl:1,2-propanediol ChCl:fructose:water ChCl:sucrose:water ChCl:maltose ChCl:glucose:water ChCl:urea	1:2 2:1 1:6 1:2 2:1 1:2 2:1:1 4:1:2 3:1 2:1:1 1:2	Microwave-assisted extraction 200 mg powder, 2 mL DES (with 25% water), 65 °C, 20 min, 700 W, 18 bar spectrometric analysis HPLC-ESI-TOF-MS	[65]
Aglycone, Demethyloeuropein, Hydroxytyrosol, Oleacein	Olive leaves, ripened olive drupes	ChCl:urea ChCl:glycerol ChCl:Lactic acid ChCl:ethylene glycol ChCl:citric acid	1:2 1:1 1:1 1:1 1:1	MAE at 100 W, S/L 1:2.5 (g/mL) with 0 or 20% water, 10 or 30 min at 80 °C, HPLC	[66]
Amentoflavone, Quercitrin, Hinokiflaveno, Myricitrin	*Platycladi Cacumen*	ChCl:levulinic acid ChCl:ethylene glycol ChCl:N,N′:dimethylurea ChCl:D-glucose Betaine:levulinic acid Betaine:ethylene glycol Betaine:1-methylurea Betaine:D-glucose L-proline:levulinic acid L-proline:glycerol L-proline:acetamide L-proline:D-glucose	1:2 1:2 1:1 1:1 1:2 1:2 1:1 1:1 1:2 1:2.5 1:1 1:1	Ultrasound 200 W, 50 °C, 30 min, centrifugation 20 min (16,200 G), suspension diluted eight times with 50% acetonitrile, HPLC-UV, The optimized DES extraction conditions: 30 min; S/L 1:4 (mg/mL) for ChCl:Levulinic Acid (90%) (1:2)	[67]
Apigenin rutinoside, Luteolin, Luteolin di-glycoside, Luteolin glucoside, Luteolin rutinoside, Oleuropein	Olive (*Olea europaea*) Leaves	Glycerol:sodium-potassium tartrate tetrahydrate:water	7:1:2	Powder leaves, 10 mL LTTM, ultrasonic power of 140 W, concentration of the LTTM in an aqueous solution (50 and 80%, w/v), S/L (1:15; 1:45 (g/mL)) and temperature (50 to 80 °C), LC-DAD-MS, total polyphenol, and flavonoid yield, antioxidant activity	[68]

Table 1. Cont.

Naringenin, Oleuropein, Caffeic acid, (±) catechin hydrate, Cinnamic acid, Gallic acid, Quercetin dehydrate, Luteolin, p-coumaric acid, Rutin hydrate, Trans-ferulic acid, Tyrosol, 3-hydroxytyrosolapigenin	olive cake, onion seed, and by product from tomato and pear canning industry	Lactic acid:glucose + 15% of water and 0.1% (v/v) formic acid	5:1	Ultrasound time (15, 35, 60 min), S/L 1:15, 1:45, 1:75 (mg/mL) and water dilution of the optimal DES (0%, 40% and 75%), temperature 40 °C. Optimization: S/L 1:75 (mg/mL) and homogenized by a vortex during 15 s. Ultrasound treatment (200 W output power, 20 kHz frequency), 60 min, 40 °C, HPLC-DAD analysis	[69]
Artemisinin	Herba Artemisiae Scopariae	[N(Me)(Oc)$_3$]Cl:ethylene glycol	1:2	S/L 1:10 (g/mL), 250 rpm, 30 °C, 15 min, HPLC-UV	[70]
		[N(Me)(Oc)$_3$]Cl:1-propanol	1:2		
		[N(Me)(Oc)$_3$]Cl:1,3-propanediol	1:2		
		[N(Me)(Oc)$_3$]Cl:glycerol	1:2		
		[N(Me)(Oc)$_3$]Cl:1-butanol	1:2		
		[N(Me)(Oc)$_3$]Cl:1,2-butanediol	1:2		
		[N(Me)(Oc)$_3$]Cl:hexyl alcohol	1:2		
		[N(Me)(Oc)$_3$]Cl:capryl alcohol	1:2		
		[N(Me)(Oc)$_3$]Cl:decyl alcohol	1:2		
		[N(Me)(Oc)$_3$]Cl:dodecyl alcohol	1:2		
		[N(Me)(Oc)$_3$]Cl:1-tetradecanol	1:2		
		[N(Me)(Oc)$_3$]Cl:cyclohexanol	1:2		
		[N(Me)(Oc)$_3$]Cl:DL-menthol	1:2		
		[N(Me)(Oc)$_3$]Cl:1-butanol:1-propanol	1:1:3, 1:2:2, 1:3:1, 1:4:0		
		[N(Me)(Oc)$_3$]Cl:1-butanol:hexyl alcohol	1:1:3, 1:2:2, 1:3:1, 1:4:0		
		[N(Me)(Oc)$_3$]Cl:1-butanol:capryl alcohol	1:1:3, 1:2:2, 1:3:1, 1:4:0		
		[N(Me)(Oc)$_3$]Cl:1-butanol:1,2-propanediol	1:1:3, 1:2:2, 1:3:1, 1:4:0		
		[N(Me)(Oc)$_3$]Cl:1-butanol:1,3-butanediol	1:1:3, 1:2:2, 1:3:1, 1:4:0		
		[N(Pr)$_4$]Br:1-butanol:1-propanol	1:1:2, 1:1.5:1.5, 1:2:1, 1:3:0		
		[N(Pr)$_4$]Br:1-butanol:hexyl alcohol	1:1:2, 1:1.5:1.5, 1:2:1, 1:3:0		
		[N(Pr)$_4$]Br:1-butanol:capryl alcohol	1:1:2, 1:1.5:1.5, 1:2:1, 1:3:0		
		[N(Pr)$_4$]Br:1-butanol:1,2-propanediol	1:1:2, 1:1.5:1.5, 1:2:1, 1:3:0		
		[N(Pr)$_4$]Br:1-butanol:1,3-butanediol	1:1:2, 1:1.5:1.5, 1:2:1, 1:3:0		
		[N(Bu)$_4$]Br:1-butanol:1-propanol	1:0.5:1.5, 1:1:1, 1:1.5:0.5, 1:2:0		
		[N(Bu)$_4$]Br:1-butanol:hexyl alcohol	1:0.5:1.5, 1:1:1, 1:1.5:0.5, 1:2:0		

Table 1. Cont.

		[N(Bu)$_4$]Br:1-butanol:capryl alcohol	1:0.5:0.5 1:1:1 1:1.5:0.5 1:2:0		
		[N(Bu)$_4$]Br:1-butanol:1,2-propanediol	1:0.5:0.5 1:1:1 1:1.5:0.5 1:2:0		
		[N(Bu)$_4$]Br:1-butanol:1,3-butanediol	1:0.5:0.5 1:1:1 1:1.5:0.5 1:2:0		
		[N(Bu)$_4$]Br:1-butanol [N(Me)(Oc)$_3$]Cl:1-butanol [N(Me)(Oc)$_3$]Cl:1-butanol:capryl alcohol	1:2 1:4 1:3:1	Extraction by air-bath shaking at 250 rpm and 30 or 60 °C, water-bath shaking at 150 rpm and 30 or 60 °C, magnetic stirring at 150 rpm and 30 or 60 °C, heating at 60 °C and 0 rpm, or ultrasonication at 200 W and 30 or 60 °C.	
		[N(Me)(Oc)$_3$]Cl:1-butanol	1:4	S/L (from 1:10 to 1:20 g/mL), ultrasonic powers (120 to 200 W), Temperature (40–60 °C), particle sizes (20 to 80 mesh), time (40–80 min)	
Astragalin, Benzoic acid, Caffeic acid, Catechinic acid, Epicatechin, Gallic acid, Gentisic acid, Hyperin, Chlorogenic acid, Quercetin, Rutin, Syringic acid, Vanillic acid	Morus alba L.	ChCl:urea ChCl:ethylene glycol ChCl:glycerol ChCl:citric acid ChCl:malic acid Betaine:levullinic acid Betaine:lactic acid Betaine:glycerol Proline:malic acid Proline:glycerol Proline:levullinic acid Proline:lactic acid	1:2 1:2 1:2 2:1 1:1 1:2 1:1 1:2 1:1 2:5 1:2 1:1	0.2 g powder, 4 mL (DES:water 3:1 v/v) sonicated at 40 °C, 30 min, centrifuged 120 rpm for 10 min HPLC-UV	[71]
Astrazon orange G, astrazon orange R, chrysoidine	Red chili peppers	ChCl:ethyl glycol ChCl:1,2 butanediol ChCl:glycerol ChCl:1,3 butanediol ChCl:1,4 butanediol	1:3 1:3 1:3 1:3 1:3	S/L 1:10 (g/mL), temperature: 30 °C, time: 20 min, and ultrasonic power: 75 W, HPLC-UV	[72]
Baicalein, Baicalin, Scutellarin, Wogonoside, Wogonin	Radix scutellatiae	ChCl:glycerol ChCl:glycol ChCl:1,2-propylene ChCl:1,4-butanediol ChCl:lactic acid ChCl:malic acid:water ChCl:glucose:water L-proline:glycerol L-proline:glucose:water L-proline:fructose:water Citric acid:fructose:water Citric acid:glucose:water	1:4 1:4 1:4 1:4 1:4 1:1:3 1:1:2 1:4 5:3:8 1:1:5 1:1:3 1:1:5	50 mg powder, 42 min, 1.2 mL DES (66.7% DES and 33.3% water), vortexed 5 min, ultrasonification 42 min, HPLC	[73]
Bergapten, Caffeoylmalic acid, Rutin, Psoralen, Psoralic acid-glucoside	Ficus carica L.	Glycerol:xylitol:D-(−)-fructose	1:3:1 1:3:2 1:3:3 2:3:1 2:3:2 2:3:3 3:3:1 3:3:2 3:3:3	DES-MAE, S/L 1:20 (g/mL), temperature 55 °C, time 10 min, microwave power 250 W, HPLC	[74]

Table 1. *Cont.*

		Glycerol:L-proline:D-(-)-fructose	3:3:3	DES-UAE, water concentration 20%, S/L 1:20 (g/mL), temperature 60 °C, time 20 min, ultrasonic power 250, 700 W, HPLC	
		ChCl:D-(+)-Galactose ChCl:L-proline ChCl:DL-malic acid ChCl:xylitol ChCl:glycerol ChCl:D-(+)-Glucose ChCl:citric acid ChCl:sucrose ChCl:D-(−)-fructose	1:1 2:1 1:1 5:2 1:1 1:1 2:1 1:1 1:1	DES-MAE, S/L 1:20 (g/mL), temperature 55 °C, time 10 min, microwave power 250 W, HPLC	
		Glycerol:L-proline:D-(−)-fructose	3:3:3	DES-MAE, water concentration 10–40%, S/L 1:5, 1:15, 1:25 (g/mL), temperature 40–80 °C, time 20–60 min, microwave power 250 W, HPLC	
Biochanin A, Daidzein, Daidzin, Genistein, Genistin	spike samples	ChCl:(+)-glucose ChCl:L(+)-tartaric acid ChCl:citric acid ChCl:saccharose ChCl:glycerine ChCl:D(+)-xylose Urea:ChCl Urea:L(+)-tartaric acid Glycerine:D(+)-glucose Glycerine:L(+)-tartaric acid Glycerine:citric acid Urea:citric acid ChCl:citric acid:glycerine ChCl:citric acid	2:1 1:1 1:1; 2:1; 1:2 2:1 1:2 1:1; 2:1 1:1 2:1 2:1 1:1 2:1 2:1 1:1:1; 2:2:1 1:1	water content 30%, S/L 1:3 (g/mL), extraction time 60 min, extraction temperature 60 °C, ultrasonic power 616 W, UHPLC-UV	[75]
		ChCl:citric acid	1:1	Central composite design: time 40–120 min, temperature 30–80 °C, ultrasonic power 264–616 W, S/L 1:3 (g/mL), 30% water content	
Caffeic acid, Catechins, Epicatechin, Protocatechuic acid	Palm bark	ChCl:ethyleneglycol ChCl:glycerol ChCl:xylitol ChCl:formic acid ChCl:citric acid ChCl:oxalic acid ChCl:malonic acid ChCl:phenol	1:1 1:1 1:1 1:1 1:1 1:1 1:1, 1:2, 1:3, 1:4, 1:5	0.5 g of the palm powder was soaked in 7.5 g of the DES and 2.5 g of H_2O in a 50 mL round-bottom flask. The mixture was refluxed at 40 °C for 6 h in a water bath for extraction. HPLC-MS	[76]
Caffeic acid, Hydroxytyrosol, Luteolin, Rutin, Vanillin Total phenolic content	Olive pomace	ChCl:citric acid ChCl:lactic acid ChCl:maltose ChCl:glycerol	1:2 1:2 1:2 1:2	Homogenate-assisted extraction 2 g olive pomace, 25 mL NADES, 30 min, 40 or 60 °C, homogenization 4000, 12,000 rpm Microwave-assisted extraction 2 g olive pomace, 25 mL NADES, 200 W, 40 or 60 °C, 30 min. Ultrasound-assisted extraction 2 g olive pomace, 25 mL NADES, 60 kHz, 280 W, 40 or 60 °C, 30 min. High hydrostatic pressure-assisted extraction 2 g olive pomace, 25 mL NADES, 300 or 600 MPa, 5 and 10 min, HPLC-DAD, spectrometric analysis	[77]

Table 1. Cont.

Caffeine, Catechin, Catechin gallate, Epicatechin, Epigallocatechin, Epicatechin-3-gallate, Epigallocatechin-3-gallate Gallatecatechin, Gallic acid, Gallocatechin	Green tea	Betaine:glycerol:glucose	4:20:1	Power irradiation 500 W, ultrasonic irradiation time 6.4–73.6 min, content of DES in the extraction solvent 24.7–100% w/w, volume of the extraction solvent per 100 mg of green tea powder 0.6–0.8 mL, LC-UV	[78]
Chlorogenic acid, (+)-catechin Gallic acid, trolox Total phenolic content, Total flavonoid content, Antioxidant activity	Coffee grounds	ChCl:urea ChCl:acetamide ChCl:glycerol ChCl:sorbitol ChCl:ethylene glycol ChCl:1,4-buatnediol ChCl:1,6-hexanediol ChCl:malonic acid ChCl:citric acid ChCl:fructose:water ChCl:xylose:water ChCl:sucrose:water ChCl:glucose:water	1:2 1:2 1:2 1:2 1:2 1:2 1:2 1:2 1:2 5:2:5 2:1:2 4:1:4 5:2:5	50 mg grounds, 0.85 mL DES irradiated at ambient temperature for 45 min, centrifuged at 12,300 g for 20 min, UPHLC-Q-TOF-MS	[79]
Chlorogenic acid, 3,4-di-O-caffeoylquinic acid, 3,5-di-O-caffeoylquinic acid, 4,5-di-O-caffeoylquinic acid	Artemisia argyi leaves	ChCl:malic acid ChCl:urea ChCl:glutaric acid ChCl:malonic acid ChCl:ethylene glycol ChCl:glycerol	1:1 1:2 1:1 1:1 1:3 1:2	20 mg powder, 1 mL solvents, ultrasonic 200 W, 40 kHz, 30 min, HPLC	[80]
		ChCl:malic acid:glutaric acid ChCl:malic acid:ethylene glycol ChCl:malic acid:glycerol ChCl:malic acid. urea ChCl:malic acid:malonic acid	2:1:1,2 2:1, 1:2:0.5 2:2:1 1:2:0.5 2:2:1 2:1:1, 1:2:1, 2:1:2 2:1:1, 1:2:1, 1:1:1, 1:1:2		
Chlorogenic acid, Quercetin-3-O-glucoside, Quercetin-O-pentoside	Juglans regia L.	ChCl:acetic acid ChCl:propionic acid ChCl:butyric acid ChCl:valeric acid ChCl:glycolic acid ChCl:lactic acid ChCl:phenylacetic acid ChCl:3-phenlacetic acid ChCl:malic acid ChCl:glutaric acid ChCl:citric acid ChCl:3-phenylpropionic acid ChCl:3-phenylbutyric acid ChCl:3-phenylvaleric acid	1:2 1:2 1:2 1:2 1:2 1:2 1:2 1:2 1:1 1:1 2:1 1:2 1:2 1:2	0.15 g powder, 5 mL DES with 20% (w/w) of water, 50 °C, 1 h, 600 rpm, HPLC	[81]
Chlorogenic acid	blueberry leaves	ChCl:ethylene glycol ChCl:glycerin ChCl:1,3-butanediol ChCl:citric acid ChCl:oxalic acid ChCl:glucose ChCl:maltose ChCl:sucrose	1:2 1:3 1:4 1:5 1:6 1:7 1:8 1:9	NPCE-DES-ATPS, temperature 60 °C, S/L 1:20 (g/mL), water concentration 20% (v/v) in DES, time 30 min and negative pressure −0.07 Pa, HPLC	[82]
		ChCl:1,3-butanediol	1:4	S/L 1:15; 1:25 (g/mL), the extraction temperature (50–70 °C) and extraction time (20–40 min), HPLC	
Cinnamyl alcohol, Rosavin, Rosin, Salidroside, Tyrosol	Rhodiola rosea L.	Lactic acid:glucose:water Lactic acid:fructose:water	6:1:6 5:1:1, 5:1:5	S/L 1:20 (g/mL), sonification 50 W, 35 kHz, 60 min, 36 °C, HPLC	[83]
Coumarin, trans-cinnamaldehyde	Cinnamomum burmannii (cinnamon bark)	ChCl:glycerol ChCl:sorbitol ChCl:xylitol ChCl:lactic acid ChCl:malic acid ChCl:citric acid Betaine:lactic acid Betaine:malic acid Betaine:citric acid	1:2 1:2 4:1 1:1 1:1 1:1 1:1 1:1 1:1	Ultrasound-assisted extraction, S/L 1:10 (g/mL)	[84]

Table 1. Cont.

Coumarin, trans-cinnamaldehyde	*Cinnamomum burmannii* (cinnamon bark), *Caesalpinia sappan* heartwoods	ChCl:glycerol		Ultrasonic extraction: 35 W, 42 Hz, S/L 1:66–1:93.75(g/mL), water content 10–80%, different ratio of glycerol to ChCl (66–20%), HPLC	[85]
Curcumin	herbal tea, turmeric drug (food supplement), turmeric powder	ChCl:phenol	1:2 1:3 1:4	VAS-DES-ELLME, HPLC, UV-VIS methodology	[86]
Epigallocatechin-3-gallate	Green tea	Betaine:glycerol:glucose	4:20:1 4:15:1 4:10:1 4:5:1	S/L 1:10 (g/mL), 45 min, irradiation power 500 W, room temperature, LC-UV	[78]
		Betaine:maltitol	4:1		
		Betaine:urea	1:2		
		Betaine:glycerol	1:1		
		Betaine:citric acid	1:1		
		Betaine:glucose	4:1		
		Betaine:maltose	4:1		
		Betaine:sucrose	4:1		
		Betaine:D-sorbitol	2:1		
		Betaine:Xylitol	4:1		
		Citric acid:Xylitol	1:1		
		Citric acid:maltitol	2:1		
		Citric acid:fructose	1:1		
		Citric acid:glycerol	1:2		
		Citric acid:glucose	1:1		
		Citric acid:maltose	2:1		
		Citric acid:sucrose	1:1		
		Citric acid:D-sorbitol	1:1		
		Glycerol:D-sorbitol	2:1		
		Glycerol:fructose	3:1		
		Glycerol:galactose	3:1		
		Glycerol:urea	1:1		
		Glycerol:glucose	3:1		
		Glycerol:maltose	3:1		
		Glycerol:sucrose	3:1		
		Glycerol:maltitol	3:1		
		Glycerol:xylitol	2:1		
		Citric acid:glycerol:glucose	1:2:1		
		Citric acid:glycerol:maltose	2:4:1		
		Citric acid:glycerol:maltitol	2:4:1		
		Betaine:glycerol:glucose	4:4:1		
		Betaine:glycerol:urea	1:1:2		
		Betaine:glycerol:maltitol	4:4:1		
		Betaine:glycerol:citric acid	1:1:1		
		Betaine:glycerol:maltose	4:4:1		
		Urea:glycerol:maltose	3:3:1		
		Urea:glycerol:maltitol	3:3:1		
		Urea:glycerol:glucose	2:2:1		
Epimedin A, Epimedin B, Epimedin C, Icariin	*Epidemium pubescens* Maxim.	ChCl:1,4-butanediol ChCl:ethylene glycol ChCl:1,2-propanediol ChCl:lactic acid ChCl:glycerol	1:5, 1:6 1:3, 1:4, 1:5, 1:6 1:4, 1:5, 1:6 1:2, 1:3, 1:6 1:1, 1:2, 1:3, 1:4, 1:5, 1:6	0.02 g of *E. pubescens* powder, 3 mL extractant, vortexing 10 min., and ultrasonic radiation at 25 °C for 20 min, and supernatant was mixed water, molar ratio 1:1 (DES/water v/v), HPLC-UV	[87]
Epimedin A, Epimedin B, Epimcedin C, Icariin, Icarisid II	Herba Epimedii	ChCl:urea ChCl:ethylene glycol ChCl:1,4-buatnediol ChCl:glycerol ChCl:glucose:water ChCl:malic acid ChCl:citric acid ChCl:lactic acid L-proline:1,2 propylene glycol L-proline:glycerol L-proline:ethylene glycol ChCl:1,2-propylene glycol	2:1 1:2, 1:3 1:3 1:4 2:1:1 1:1 1:1 1:2 1:3 1:4 1:3, 1:4, 1:5, 1:6 1:2, 1:3, 1:4, 1:5, 1:6	0.2 g powder, 4 mL solvent (DES:water 7:3, v/v), mixed by vortex 5 min, ultrasonic extractionat room temperature for 45 min. HPLC	[88]
Flavonoids	*Carthamus tinctorius*	ChCl:oxalic acid ChCl:ethylene glycol ChCl:1,3-butanediol ChCl:1,6-hexanediol	1:1	Ultrasonic treatment: 0.5 g powder, solvents 10–35 mL, 45 °C, 20 min, 150 W Other ultrasonic treatment: different conditions, change the parameters: 10–60 min, 60–240 W, 25–45 °C Spectrometric analysis	[89]

Table 1. Cont.

Compound	Source	DES	Ratio	Method	Ref.
Ginkgolide A	Ginkgo biloba	ChCl:glycerol	1:2	UAE, 70% (w/w) aqueous solution at 100 W and 25 °C for 10 min, S/L 1:15 (g/mL), colorimetric method	[90]
		ChCl:ethylene glycol	1:2		
		Xylitol:levulinic acid	1:1		
		1,2-propanediol:levulinic acid	1:1		
		1,3-butanediol:levulinic acid	1:1		
		Betaine:ethylene glycol	1:3		
		Betaine:levulinic acid	1:3		
		Betaine:glycerol	1:2		
		ChCl:urea	1:2		
		ChCl:levulinic acid	1:2		
		ChCl:glycolic acid	1:1		
		ChCl:glutaric acid	1:1		
		ChCl:D-sorbitol	1:1		
		ChCl:xylitol	1:1		
		ChCl:1,3-butanediol	1:3		
		ChCl:1,2-propanediol	1:2		
		Betaine:ethylene glycol + water	6:4	Magnetic stirring at 45 °C for 20 min, colorimetric method	
				UAE at 45 °C and 100 W for 20 min, colorimetric method	
Bilobalide, Ginkgolide A, Ginkgolide B, Ginkgolide C	Ginkgo biloba	Betaine:ethylene glycol + water	1:2	DES containing water 0–100% w/w, S/L 1:15 (g/mL) with ultrasound at 100 W and 25 °C for 10 min., colorimetric and HPLC-ELSD method	[90]
		ChCl:urea + water	1:2		
		Betaine:ethylene glycol + water	1:2	Water 40% w/w, S/L 1:15 (g/mL) with ultrasound at varied temperature (25–60 °C) and 100 W for 10 min., colorimetric and HPLC-ELSD method	
		ChCl:urea + water	1:2		
		Betaine:ethylene glycol + water	1:3	Water 40% w/w, S/L (1:7.5, 1:10, 1:12.5, 1:15, 1:20, 1:30, and 1:50 (g/mL)), with ultrasound at 45 °C and 100 W for 10 min., colorimetric and HPLC-ELSD method	
		Betaine:ethylene glycol + water	1:3	S/L 1:10 (g/mL) with ultrasound at 45 °C and 100 w for different time 5–40 min., colorimetric and HPLC-ELSD method	
Glycyram, Licuroside	Glycyrrhizae roots	Sorbitol:malic acid:water	1:1:3	S/L 1:10 (g/mL), 24 h, 25 °C, RP HPLC	[91]
Hespederin	Mandarin peels	ChCl:acetamide	1:2	50 mg powder, 1 mL solvent (DES with 20% (v/v) water), stirring 50 °C for 30 min, HPLC-DAD	[92]
		ChCl:1,4-butanediol	1:2		
		ChCl:citric acid	1:1		
		ChCl:ethylene glycol	1:1		
		ChCl:glycerol	1:2		
		ChCl:lactic acid	1:1		
		ChCl:levulinic acid	1:1		
		ChCl:malonic acid	1:1		
		ChCl:malic acid	1:1		
		ChCl:N-methyl urea	1:3		
		ChCl:oxalic acid	1:1		
		ChCl:sorbitol	1:1		
		ChCl:urea	1:1		
		ChCl:thiourea	1:1		
		ChCl:xylitol	1:1		
Indole-3-acetic acid, 1-naphtaleneacetic acid	Fruit juice	Benzyltriethylammonium chloride:thymol	1:4	Fruit juice samples diluted with in ratio 1:10, VA-DES-DLLME, HPLC	[93]
		[N(Me)(Oc)$_3$]Cl:butanol	1:4		
		[N(Me)(Oc)$_3$]Cl:isoamyl alcohol	1:4		
		[N(Me)(Oc)$_3$]Cl:octanol	1:4		
Levofloxacin	Green bean	Betaine:ethyleneglycol:water	1:2:1	SPE-HPLC	[94]
Lignin content in delignified biomass	oil palm biomass residues, empty fruit bunch	Malic acid:ChCl-water	2:4:2 (L-malic acid)	S/L 1:20 (w/w), 85 °C, overnight	[95]
		Malic acid:ChCl-water	2:4:2 (cactus)		
		Malic acid:monosodium glutamate:water	3:1:5 (L-malic acid)		
		Malic acid:monosodium glutamate:water	3:1:5 (cactus)		

Table 1. Cont.

Compound	Source	DES	Molar ratio	Conditions	Ref.
Oxyresveratrol	*Morus alba* Roots	Urea:glycerin	1:1, 1:2, 1:3	1 g powder, 20 mL NADES, ultrasonic treatment 10, 15, 20 min, HPLC	[96]
Pectin	pomelo (*Citrus grandis* (L.) Osbeck) peels	Lactic acid:glucose:water Lactic acid:glycine Lactic acid:glucose Lactic acid:Glycine:water	6:1:6, 5:1:3 9:1 5:1 3:1:3	S/L 1:20 (g/mL), 60 min, 50 °C, 500 rpm S/L 1:20 (g/mL), 45 min, 70 °C, 55 rpm	[97]
Phlorotannin content	Brown algae: *Fucus vesiculous* L., *Ascophyllum nodosum* L.	ChCl:lactic acid ChCl:malic acid. Water Glucose:lactic acid:water Betaine:malic acid:water Betaine:lactic acid:water Betaine:malic acid:glucose Betaine:glycerin:glucose	1:1, 1:2, 1:3 1:1:1, 2:1:1 1:5:3 1:1:1 1:2:1 1:1:1 1:5:1	20 g algae, 100 mL solvents (pure DES or with water content 50–70%), 120 min, 50 °C, spectrometric analysis	[98]
Polyprenyl acetates	*Ginkgo biloba* leaves	[N(Me)(Oc)$_3$]Cl:hexanoic acid [N(Me)(Oc)$_3$]Cl:octanoic acid [N(Me)(Oc)$_3$]Cl:capric acid [N(Me)(Oc)$_3$]Cl:lauric acid [N(Me)(Oc)$_3$]Cl:myristic acid [N(Me)(Oc)$_3$]Cl:palmitic acid [N(Me)(Oc)$_3$]Cl:octadecenoic acid [N(Me)(Oc)$_3$]Cl:ricinoleic acid [N(Me)(Oc)$_3$]Cl:1-propanol [N(Me)(Oc)$_3$]Cl:1-butanol [N(Me)(Oc)$_3$]Cl:hexyl alcohol [N(Me)(Oc)$_3$]Cl:capryl alcohol [N(Me)(Oc)$_3$]Cl:decyl alcohol [N(Me)(Oc)$_3$]Cl:dodecyl alcohol [N(Me)(Oc)$_3$]Cl:1-tetradecanol [N(Me)(Oc)$_3$]Cl:1-hexadecanol [N(Me)(Oc)$_3$]Cl:cyclohexanol [N(Me)(Oc)$_3$]Cl:DL-menthol [N(Me)(Oc)$_3$]Cl:capryl alcohol:octylic acid	1:2 1:2 1:2 1:2 1:1 1:1 1:2 1:2 1:2 1:2 1:2 1:2 1:2 1:1 1:2 1:2 1:2 1:2 1:2:3	80 mg of Ginkgo biloba leaves powder was extracted with 0.80 mL of the DES by heating at 60 °C and 0 rpm, stirring at 150 rpm and 25 or 60 °C, water-bath shaking at 150 rpm and 25 or 60 °C, air-bath shaking at 250 rpm and 25 or 60 °C, ultrasonic treating at 200 W and 25 or 60 °C, HPLC-DAD	[99]
Proanthocyanidin	*Gingko biloba* leaves	ChCl:glycerol ChCl:ethylene glycol ChCl:propylene glycol ChCl:1,3-buatnediol ChCl:sorbitol ChCl:xylitol ChCl:1,5-pentanedioic acid ChCl:glycolic acid ChCl:malonic acid ChCl:malic acid ChCl:levulinic acid ChCl:lactic acid ChCl:citric acid ChCl:tartaric acid ChCl:urea ChCl:oxalic acid	1:2 1:2 1:2 1:3 1:1 1:1 1:1 1:1 1:1 1:1 1:1 1:2 1:1 1:1 2:1 1:1	100 mg powder, 1 mL DES with 30% water, shaking at 250 rpm, 25 °C, 5 min, centrifugation at 10,000 rpm for 10 min, spectrometric analysis	[100]
Protein	Brewer's spent grain	Sodium formate:urea Potassium acetate:urea Sodium acetate:urea	1:2, 1:3 1:2, 1:3 1:2, 1:3	90 wt% carboxylate salt—urea DESs at 10 wt% consistency, 90 °C and time 4 h	[101]
		ChCl:urea Sodium acetate:urea	1:2 1:2	consistency 5 or 10% wt, defat samples, 80 °C, 20 h extraction	
Protein	Bamboo shoots and sheath	ChCl:levulinic acid	1:2, 1:3, 1:4, 1:5, 1:6	S/L 1:30 to 1:60; Temperature 20–40 °C, water content 5–30%	[102]
Quercetin	Ginkgo biloba	ChCl:glycerol ChCl:ethylene glycol ChCl:1,4-butanediol	1:2, 1:3, 1:4, 1:5 1:2, 1:3, 1:4, 1:5 1:2, 1:3, 1:4, 1:5	powder (2.0 g) was dissolved in 40 mL methanol, ultrasonic treated (60 W), 30 min, HPLC	[103]
Quercetin, Quercetin-3-O-glucoside, Isorhamnetin, Kaempferol, Rutin	Sea buckthorn leaves	ChCl:citric acid ChCl:malic acid ChCl:lactic acid ChCl:ethylene glycol ChCl:1,3-butanediol ChCl:1,4-butanediol ChCl:1,6-hexanediol ChCl:1,2-propanediol ChCl:glycerol ChCl:glucose ChCl:fructose ChCl:sucrose	1:1 1:1 1:1 1:1 1:1 1:1 1:1 1:1 1:1 1:1 1:1 1:1	microwave-assisted extraction, 1.0 g of leaves, 20 mL DES with 20% (v/v) water, 600 W, 17 min, HPLC	[104]

Table 1. Cont.

Quercetin, Isorhamnetin, Kaempferol, Naringenin	Pollen Typhae	ChCl:1,4-butanediol ChCl:glucose ChCl:glycerol ChCl:1,4-buatnediol:glycerol L-proline:glycerol ChCl:lactic acid ChCl:ethylene glycol ChCl:1,2-propanediol	1:4 1:4 1:4 1:2:2 4:11 1:4 1:4 1:4	100 mg powder, 2 mL DES, vortexed 5 min, and ultrasonic irradiation 35 min, centrifugation at 4200 rpm for 25 min, HPLC-UV	[105]
Quercetin, Isorhamnetin, Naringenin, Kaempferol, Myrecetin,	Flos Sophorae	ChCl:malic acid ChCl:citric acid ChCl:malonic acid ChCl:methylurea ChCl:urea ChCl:N,N-dimethylurea ChCl:1,3-butanediol ChCl:ethylene glycol ChCl:glycerol	1:1, 1:3 1:1, 1:3 1:1, 1:3 1:1, 1:3 1:1, 1:3 1:1, 1:3 1:1, 1:3 1:1, 1:3 1:1, 1:3	200 mg powder, 1 mL DES, and short homogenization, AP/MALDI-MS	[106]
Rosmarinic acid, Rutin	Satureja montana L.	ChCl:urea ChCl:sorbitol ChCl:1,4-butanediol ChCl:lactic acid ChCl:levulinic acid	1:2 1:1 1:2 1:2 1:2	50 mg leaves, 1 mL solvents (DES + water (10, 30, 50% of water, v/v), stirring at 1500 rpm, 30, 50, 70 °C for 60 min, HPLC	[107]
Rutin	tarary buckwheat hull	ChCl:1,2-propanediol ChCl:glycerol ChCl:glucose ChCl:sucrose ChCl:xylitol ChCl:sorbitol Glycerol:L-proline Glycerol:L-alanine Glycerol:L-histidine Glycerol:L-threonine Glycerol:L-lysine Glycerol:L-arginine	1:1 1:1 2:5 1:1 1:2 2:5 3:1 3:1 3:1 3:1 4.5:1 4.5:1	40 mg of tartary buckwheat hull powder, 1.0 mL solvent, 40 °C, 60 min, UAE power 200 W,	[108]
Total flavonoids and polyphenols, and total polyphenols at saturation tentative identity: Apigenin C-glycoside, Chlorogenic acid, Quercetin diglycoside, Quercetin glycoside, Quercetin glycoside derivative, Quercetin rhamnoside derivative, Quercetin malonylglycoside derivative, Kaempferol glycoside derivative, Kaempferol malonylglucoside, Multiflorin B	Moringa oleifera Lam. leaves	Glycerol:sodium acetate	4:1 5:1 6:1	2.5 g of lyophilized leaves was mixed with 50 mL of aqueous LTTM mixture and stirred at 600 rpm for 180 min, at 50 °C, LC-DAD-MS, total flavonoids and polyphenols	[109]
Total phenolic and anthocyanin content	Hibiscus sabdariffa	Citric acid:glycerol Citric acid:ethylene glycol	1:4 1:4	Microwave-assisted extraction, 60 to 150 s, power 250, 350, 450, 550, 600 W, spectrometric analysis antioxidant activity determined	[110]
Total polyphenolic and flavonoid contents, Chlorogenic acid, chlorogenic acid isomer, Quercetin glucoside, quercetin malonylglycoside derivate, Kaempferol glucoside, kaempferol malonylglucoside, Multiflorin B, Neochlorogenic acid	Moringa oleifera L.	Glycerol:nicotinamide	5:1	Ultrasonic pretreatment: 0.57 g plant, 20 mL solvent (70% w/v aqueous solution), 50 Hz, 550 W, acoustic energy density 78.6 W/L, 23 °C, 5–40 min Batch stirred-tank extraction: 0.57 g plant, 20 mL solvent (70% w/v aqueous solution), 50 °C, 150 min, spectrometric analysis, HPLC antiradical activity, reducing power	[111]

Table 1. Cont.

Total phenolic content	Ruta graveolens L.	ChCl:citric acid	2:1	50 mg leaves, 1 mL solvent with different content of water (10–30%), stirring at time 30, 52, 60, 90 min, 30, 50, 70 °C, RP-HPLC	[112]
Total phenolic content	Spruce bark	ChCl:lactic acid:water	1:2:0.96	0.5 g powder, 10 mL DESs, stirring at 60 °C for 2 h, spectrometric analysis antioxidant activity determined	[113]
		ChCl:lactic acid:water	1:3:0.97		
		ChCl:lactic acid:water	1:4:0.99		
		ChCl:lactic acid:water	1:5:0.98		
		ChCl:lactic acid:1,3-propanediol:water	1:1:1:0.92		
		ChCl:lactic acid:1,3-propanediol:water	1:2:1:0.95		
		ChCl:lactic acid:1,3-propanediol:water	1:3:1:0.91		
		ChCl:lactic acid:1,3-propanediol:water	1:4:1:0.92		
		ChCl:lactic acid:1,3-propanediol:water	1:5:1:0.91		
		ChCl:lactic acid:1,3-propanediol:water	1:1:1:0.93		
		ChCl:lactic acid:1,3-propanediol:water	1:2:1:0.92		
		ChCl:lactic acid:1,3-propanediol:water	1:3:1:1		
		ChCl:lactic acid:1,3-propanediol:water	1:4:1:1		
		ChCl:lactic acid:1,3-propanediol:water	1:5:1:1		
		ChCl:lactic acid:1,3-butanediol:water	1:1:1:0.96		
		ChCl:lactic acid:1,3-butanediol:water	1:2:1:0.92		
		ChCl:lactic acid:1,3-butanediol:water	1:3:1:0.92		
		ChCl:lactic acid:1,3-butanediol:water	1:4:1:0.91		
		ChCl:lactic acid:1,3-butanediol:water	1:5:1:0.91		
		ChCl:lactic acid:1,3-butanediol:water	1:1:1:0.87		
		ChCl:lactic acid:1,3-butanediol:water	1:2:1:0.98		
		ChCl:lactic acid:1,3-butanediol:water	1:3:1:0.90		
		ChCl:lactic acid:1,3-butanediol:water	1:4:1:0.90		
		ChCl:lactic acid:1,3-butanediol:water	1:5:1:0.96		
		ChCl:lactic acid:1,4-butanediol:water			
		ChCl:lactic acid:1,4-butanediol:water			
		ChCl:lactic acid:1,4-butanediol:water			
		ChCl:lactic acid:1,4-butanediol:water			
		ChCl:lactic acid:1,4-butanediol:water			
		ChCl:lactic acid:1,5-pentanediol:water			
		ChCl:lactic acid:1,5-pentanediol:water			
		ChCl:lactic acid:1,5-pentanediol:water			
		ChCl:lactic acid:1,5-pentanediol:water			
		ChCl:lactic acid:1,5-pentanediol:water			
Total phenolic content, boldine, 9 alkaloids and 22 phenolic compounds	Peumus boldus leaves	ChCl:1,2-propanediol	1:3	Plant 0.1 g, 10 mL NADES (80% aqueous solution), vortexed 30 s, stirring extraction 60 °C, 50 min, 340 rpm Ultrasound extraction: room temperature, 20 min, 140 W, 37 Hz HPLC-PDA-ESI-IT/MS, HPLC-ESI-QTOF-MS	[114]
		ChCl:glycerol	1:2		
		ChCl:lactic acid	1:2		
		ChCl:levulinic acid	1:1		
		L-proline:citric acid	1:2		
		L-proline:oxalic acid	1:1		
		L-proline:levulinic acid	1:1		
Total polyphenolic and flavonoid contents	Thyme (Coridothymus capitatus, Thymus vulgaris), Oregano (Origanum vulgare hirtum), Greek sage (Salvia fruticosa), Sage (Salvia officinalis)	Lactic acid:nicotinamide	7:1	0.57 g of dried plant material, added 20 mL solvent, S/L 1:30 (g/mL), treated UAE, 37 Hz, 140 W, extraction time 60 min, 55 °C, extraction by aqueous DES solutions (75% v/v), other extraction β-cyclodextrin was added to the mixture (1.5% w/v), antiradical activities, reducing power determined	[115]
		Lactic acid:ChCl	7:1		
		Lactic acid:sosium acetate	7:1		
		Lactic acid:ammonium acetate	7:1		
		Lactic acid:glycine	7:1		
		Lactic acid:L-alanine:	7:1		
Total polyphenolic and flavonoid contents Chlorogenic acid, Di-caffeoylquinic acid, di-p-coumaroylquinic acid derivate, Isoquercetin, Quercetin, Narcissin, neochlorogenic acid, rutin	Sambucus nigra flowers	Lactic acid:glycín	5:1, 7:1, 9:1, 11.1, 13:1	Ultrasonic pretreatment: 0.57 g plant, 20 mL solvent (70% w/v aqueous solution), 50 Hz, 550 W, acoustic energy density 75.3 W/L, 22 °C, 5–40 min Batch stirred-tank extraction: 0.57 g plant, 20 mL solvent (70% w/v aqueous solution), 50 °C, 150 min, spectrometric analysis, HPLC-DAD, LC-DAD-MS antiradical activities, reducing power determined	[116]

Table 1. Cont.

Vanillin	Vanilla pods (Vanilla planifolia)	Betaine:citric acid Lactic acid:1,2-propandiol Lactic acid:fructose Fructose:glucose		15 mg of vanilla pods were extracted 1 mL NADES, water content (90:10, 75:25, 60:40, 40:60 (v/v)), HPLC-DAD	[117]
Vanillin	Vanilla pods (Vanilla planifolia)	ChCl:citric acid:water ChCl:malic acid:water ChCl:glycerol Fructose:glucose:water Malic acid:glucose:water Betaine:sucrose:water Betaine:citric acid:water Betaine:malic acid:glucose:water Citric acid:fructose:glucose:water Malic acid:glucose:fructose:water L-Serine:malic acid:water B-alanine:citric acid:water Lactic acid:1,2-propanediol Lactic acid:fructose	1:1:6 1:1:6 1:1 1:1:6 1:1:6 2:1:6 1:1:6 1:1:1:9 1:1:1:9 1:1:1:9 1:1:6 1:1:6 1:1 5:1	50 mg of vanilla pods were extracted 50 °C, 1 h, HPLC-DAD	[117]

Choline chloride—ChCl; methyl trioctyl ammonium chloride—[N(Me)(Oc)$_3$]Cl; tetrabuthylammonium bromide—[N(Bu)$_4$]Br, tetrapropylammonium bromide—[N(Pr)$_4$]Br.

7. Assessing the Main Opportunities of Using Phytomass for Extraction of High Value-Added Components by Deep Eutectic Solvents

Today's level of chemical synthesis makes it possible, in principle, to prepare any chemical compound at the laboratory and industrial scale. The properties of chemical compounds, including those of biological origin, do not depend on the method of their preparation. The advantage of compounds and substances isolated from natural sources over synthetically prepared ones lies in several factors. One of them is the cost related to their obtaining, which in the case of renewable natural resources can be significantly lower. Possible causes of different therapeutic effects are given in the section "Therapeutic effects of substances extracted from phytomass". As in most cases of groups of related compounds (polyphenols, flavonoids, etc.) being separated from phytomass by DESs, we will use the term value-added substances in the following text. In addition, the most suitable case is a situation where entire extracts (i.e., DESs – extracted substance) can be used directly without their prior separation.

When the impact of obtaining substances from biological materials on the environment is also taken into account, it is logical that methods with a minimal adverse effect will be preferred. Thus, the extraction methods will preferably be those using green solvents, including DESs. Discussing the properties of substances referred to as value-added ones, we will focus on those exhibiting therapeutic effects and applied in the food sector.

In studies published mainly during the few last years, numerous value-added substances were obtained using various extraction techniques and green solvents. The attention was focused predominantly on phytomass containing a relevant amount of substances, denoted usually as bioactive compounds. Taking the potential of renewable phytomass processing into account, the investigation of extraction was directed to isolation of such substances in the highest possible yield. The value-added substances isolated from phytomass, its waste, and food waste can be classified based on their biological properties, structural or chemical class of compounds, actual or potential applicability, etc. Many of the value-added substances can be isolated from different sources using various extraction techniques and green solvents. The mentioned factors essentially make it impossible to unambiguously classify the extracted value-added compounds. The spectrum of the properties of these compounds is really wide (anticoagulative, anti-inflammatory, antioxidant, hepatoprotective, antihypertensive, antitumor, antimicrobial, anticancer, antidiarrheal, antiallergic, antiatherosclerotic, estrogenic, insecticidal, antimutagenic, pharmacokinetic, antiproliferative, neuroprotective, antiangiogenetic, antagonist, and others) and, therefore, their application is possible in different areas [62,118–120].

The most important potential use of these compounds isolated using DESs includes pharmaceutical and biomedical applications, and last but not least, application in the food industry such as additives and functional substances, nutraceuticals used in the food industry and to enhance food quality.

In the following, we will focus on individual types of extracted substances based on published data. Most of them are phenolic compounds, where groups of substances (total polyphenols), their subgroups (flavonoids), or even individual compounds (rutin) have been isolated and determined.

From the following data, it is clear that the extraction experiments were performed at a laboratory scale. The yield parameters of large-capacity extractions may vary due to different operating conditions.

7.1. Total Polyphenols

Polyphenols—organic compounds found in plants—include more than 8000 compounds. Particular attention is devoted mainly to curcumin, resveratrol, catechins, anthocyanins, and flavonoids. Interest in these substances stems from their vital role in health through the regulation of metabolism, weight, chronic disease, and cell proliferation. In vitro and in vivo studies indicate that polyphenols have antioxidant and anti-inflammatory properties that could have preventive and/or therapeutic effects for cardiovascular disease, neurodegenerative disorders, cancer, and obesity [121]. Thus, it is natural that polyphenols are often separated not only by DESs but also by other solvents.

A comparison of the extraction efficiency of green and other solvents, polyphenols, and flavonoids determined the antioxidant activity and reducing power from Olive (*Olea europaea*) leaves, using five different solvents (water, 60% methanol, 60% ethanol, 9% (w/v) aqueous glycerol, and 50% (v/v) DES). In accordance with the results by Georgantzi et al. [115], it was found that to efficiently extract flavonoids, conventional solvents (methanol, ethanol, or aqueous glycerol) should be preferred. The antioxidant activity and reducing power was lowest when working with DESs, which was ascribed to a lower amount of extracted flavonoids. The same conclusion can be found also in an older paper published by Lee et al. [122]. Glycerol and sodium acetate in various molar ratios, acting as DESs [109], were used for extraction of *Moringa oleifera Lam.* leaves. The content of extracted polyphenols and flavonoids, antioxidant activity, and reducing power were compared to results obtained using ethanol (80% v/v). In spite of a higher yield of polyphenols (51.69 mg GAE/g dw (dry weight)) and flavonoids (16.48 mg RtE/g dw) against the conventional solvent (30.05 mg GAE/g dw; 13.76 mg RtE/g dw), antioxidant activity of DESs was lower. This disagreement is explained by the authors as a result of synergism or antagonism among the polyphenolic constituents. This conclusion was supported also by Philippi et al. [123].

UAE-DES extraction of olive cake, onion seed, tomato, and pear by lactic acid:glucose (5:1), 15% water, and 0.1% (v/v) formic acid was evaluated for different byproducts [69] with the aim to determine the yield of various phenolic compounds (Table 1). As a result of the optimization, lyophilized material and DES were homogenized by a vortex during 15 s and the suspensions were processed by ultrasound (200 W, 20 kHz) for 60 min at 40 °C.

Ruta graveolens L. as a rich source of phytochemicals has been used to extract polyphenolic substances using ChCl and citric acid (2:1). The extracts obtained at 30 °C with 20% water content and at a time of 90 min with DES extract content (13.3 g/mL) reached the highest polyphenol content 38.24 mg GAE/g dry matter and the highest antioxidation activity 72.53%. The extracts had antibacterial properties, especially against Gram-negative bacteria *P. aeriginosa* [112].

Extraction of different types of phenolic substances [80] using six binary DESs was applied to a traditional Chinese medicinal plant of the genus Artemisia (*Artemisia argyi*). In addition, the effect of ternary DESs, which contained ChCl, malic acid, and a third component (urea, ethylene glycol, glycerol, glutaric acid, and malonic acid), was investigated. Ternary DESs containing ChCl, malic acid, and urea (2:1:2) showed higher extraction yields for phenolic acids compared to conventional organic solvents and other DESs. The optimal conditions for achieving the highest yield of phenolic compounds for this system were: extraction time 23.5 min; liquid to substrate ratio 57.5 mL/g dry plant material; water content 54%.

The extraction of polyphenols from *M. oleifera* leaves with a new type of DES, which contained glycerol and nicotinamide, was performed by ultrasonic pretreatment. The result was the optimization of the process in which the highest yield of polyphenols (82.87 mg GAE/g dry biomass) was obtained after 30 min of ultrasonic pretreatment [111]. Ultrasonic pretreatment was also used by Kaltsa et al. [116]

who extracted polyphenols from the flowers of the black elder (*Sambucus nigra*). In this case, the effect of ultrasound was confirmed, which ensured a higher yield of the polyphenols of interest using the DES containing lactic acid and glycine.

DESs containing water based on ChCl with lactic acid, 1,3-propanediol, 1,3-butanediol, 1,4-butanediol, and 1,5-pentanediol, with different molar ratios, were used as extractants for the extraction of polyphenols from spruce bark [120]. The content of polyphenols in the extracts ranged from 177.6 to 596.2 mg of GAE per 100 g of dry bark. In addition to the content of polyphenols, antioxidant activity was also evaluated. Differences in radical scavenging activity (RSA) indicate that each DES preferentially dissolved a different type of extractant with a different reactivity to DPPH$^{\bullet}$. The RSA values of the extracts (i.e., containing the DES system and the extracted substances) ranged from 81.4% to 95%. Lower antioxidant activity (RSA 86.4%) was observed for extracts obtained with ChCl:lactic acid:water (1:2:0.96), and for the system containing ChCl:lactic acid and various diols in a molar ratio of 1:1:1, namely 82.4% for 1,3-propanediol; 84.2% for 1,3-butanediol; 85.4% for 1,4-butanediol; 81.4% for 1,5-pentanediol. ChCl:lactic acid:1,3-butanediol:water extracts (1:5:1:1) exhibited the highest antioxidant activity (RSA 95%), and this extract also had the highest polyphenol content (596.2 mg GAE/100 g dry bark).

Bioactive substances such as trans-cinnamaldehyde and coumarin were extracted using DESs by Sakti et al. [85] and Aryati et al. [84]. Cinnamon bark (*Cinnamomum burmannii*) was used as a source of these substances. Both studies examined the effect of ultrasound extraction in combination with DESs. Sakti et al. [85] applied ChCl and glycerol; Aryati et al. [84] ChCl (six kinds) and betaine (three kinds). It has been shown that higher yields of the extracted substances can be reached under suitable conditions using the DESs than by application of conventional methods such as reflux, Soxhlet, or maceration using an organic solvent (96% ethanol) [84].

7.2. Phlorotannins

Phlorotannins are a class of polyphenol compounds exhibiting a variety of biological activities, and are used as antifungal, antimicrobial, antioxidant, anticoagulant, antiallergic, antihyperlipidemic, algicidal, and enzyme-inhibitory agents [124–126]. Obluchinskaya et al. [98] used DESs to extract phlorotannin from brown algae (*Fucus vesculosus* L., *Ascophyllum nodosum* (L.) Le Jolis). The extraction efficiency of polyphenols is evaluated using 10 DESs containing ChCl, betaine, and glucose in different molar ratios. The extraction was performed as maceration at 120 min, 50 °C with a phytomass to extractant ratio of 1:5. When 50–70% aqueous solutions of DESs (ChCl with addition of lactic or malic acid and also malic acid and betaine) were applied, the maximum extraction efficiency of phlorotannin reached 60–72%.

7.3. Flavonoids

Flavonoids are a class of polyphenolic plant and fungus secondary metabolites. Flavonoids are of interest due to their antioxidant properties antioxidative, anti-inflammatory, antimutagenic, and anticarcinogenic properties coupled with their capacity to modulate key cellular enzyme functions [127,128]. Of all flavonoids, anthocyanins, quercetin, kaempferol, rutin, and their derivatives are the most studied in terms of DESs extraction. Quercetin and derivates are plant flavonoid pigments and have a wide range of biological actions including anticarcinogenic, anti-inflammatory and antiviral activities, as well as attenuating lipid peroxidation, platelet aggregation, and capillary permeability [129,130]. The extraction of flavonoids such as quercetin, kaempferol, naringenin, and isorhamnetin from *Pollen Typhae* by ultrasound-assisted deep eutectic solvents extraction was realized by Meng et al. [105]. DESs showed greater extraction efficiency of flavonoids comparing with conventional solvents such as water, ethanol, methanol, and 75% of aqueous ethanol. The highest extraction efficiency was achieved by application of ChCl and 1,2-propanediol (1:4) with water (30%).

This conclusion was supported also by Cui et al. [104]. The extraction of quercetin, quercetin-3-O-glucoside isorhamnetin, kaempferol, and rutin from sea buckthorn leaves using the selected 12 DESs was more efficient than that performed by 70% ethanol. Target flavonoids reached the yield of 20.82 mg/g for optimized conditions [104].

The genus Epimedium is rich in terms of flavonoids, of which icariin, epimedin A, epimedin B, and epimedin C are known especially to be biologically active, such as antitumor, an immunoenhancing effect, and improvement in the function of the cardiovascular, nervous, and immune systems [131,132]. ChCl in combination with 1,2-propanediol, 1,4-butanediol, glycerol, or lactic acid in various molar ratios has been used to extract substances such as epimedin A, B, C, and icariin from the Chinese medicinal herb *Epidemium pubescens Maxim*. The highest extraction efficiency of prenylated flavonol glycosides was achieved using the DES composed of ChCl and lactic acid (1:2) [87]. Kulturbas and coworkers [110] used microwave extraction of Sudanese hibiscus (*Hibiscus sabdariffa*) by the DESs containing citric acid and glycerol or ethylene glycol. Parameters such as yield of polyphenols, anthocyanates, and antioxidant activity were monitored. A system containing citric acid and ethylene glycol (35 mL DES, including 50% water, microwave power 550 W) was evaluated as the most suitable and efficient. Guo et al. [88] applied ultrasonic extraction in combination with DES and evaluated the extraction efficiency of substances such as: epimedin A, epimedin B, epimedin C, icariin, and icariside II. A screening evaluation of 12 types of DES for the extraction of the mentioned substances from a plant known as Chinese viagra (*Herba Eminedii*) was performed. Based on the screening evaluation, the DES containing L-proline and ethylene glycol in a molar ratio (1:4) was selected and used in the planned experiment. Optimal conditions for extraction of flavonoids are: 0.2 g of substrate in powder form and ultrasonic extraction for 45 min using 4.00 mL of a 70% aqueous solution of the mentioned DES. Comparing the extraction of icariin with the traditional method described in the *Chinese Pharmacopoeia* (2015 edition), solvent consumption was reduced by 80% and extraction time was shortened by 25%.

7.4. Catechins

Catechins (flavan-3-ols) belong to the group of polyphenols, and with other catechin flavonoids have antioxidant and anti-inflammatory properties, and affect the molecular mechanisms involved in angiogenesis, extracellular matrix degradation, the regulation of cell death, and multidrug resistance in cancers and related disorders [133].

Jeong et al. [78] tested the impact of 42 DESs on the ultrasonic-assisted extraction of catechin from green tea (Table 1) and compared extraction efficiency with that obtained using water, methanol, ethanol, 70% methanol, and 70% ethanol. The authors found that all the DESs are more suitable to extract epigallocatechin-3-gallate than water or ethanol. Moreover, they verified possibilities to apply ternary systems composed of betaine:glycerol:glucose in various molar ratios. The performed screening evaluation and comparison with water, methanol, ethanol, 70% methanol, or ethanol led to the conclusion that the ternary systems exhibited a higher extraction power. When applying the optimized system with the following variables: ultrasonic irradiation time 6.4–73.6 min, content of DES in the extraction solvent 24.7–100% w/w, volume of the extraction solvent per 100 mg of green tea powder 0.6–0.8 mL, and the mentioned ternary 4:20:1 system, the maximum yield of epigallocatechin-3-gallate was 102.3 mg/g, and that of total catechins 217.7 mg/g (optimal conditions: 81% DES, room temperature, 6.5 min) was reached. UAE with DES was identified as being the best system for catechins compounds extraction, followed by stirring (50% ethanol, room temperature, 150 min—165.9 mg/g); heating (water, 80 °C, 30 min—101.5 mg/g); UAE (water, 60 °C, 40 min—100.7 mg/g) and heating + stirring (water, 80 °C, 40 min—93.3 mg/g). It can be concluded that Jeong et al. [78] documented a possibility of the use of green solvents to extract catechin substances, while combination with UAE allows the reaching of high extraction yield in a relatively short time without the necessity to heat the solvent. Fu et al. [76] compared the effect of methanol and eight different types of DESs (Table 1) following the extraction of polyphenols from palm bark. They also documented, in accordance with Jeong et al. [78], that extraction using the selected DESs was more efficient than that performed by methanol.

Fu et al. [76] pre-treated palm bark samples by ChCl with ethyleneglycol, glycerol, xylitol, phenol, formic acid, citric acid, oxalic acid, and malonic acid. DES-modified adsorbent and DES-effluent was used for solid-phase extraction. The extractability of bioactive compounds such as protocatechuic acid, catechins, epicatechin, and caffeic acid were monitored. Results showed that the eco-friendly extraction has high potential degree to be introduced to the area of new analytical and extraction methods. Georgantzi et al. [115] found that when using DESs as extractants, higher yields of extracted polyphenols were reached than those using water in all cases (Table 1). An analogous result was obtained when applying 60% ethanol. Comparing extraction yield reached by DESs enriched by 1.5% *w/v* cyclodextrin showed that in majority cases the extractant containing cyclodextrin was more effective than water or a water-ethanol mixture. On the other hand, to extract flavonoids from various plants, the results obtained using water or water-ethanol mixture as extractants led to a higher yield than those obtained by pure or cyclodextrin-enriched DESs. The authors investigated also influence of agents to antiradical activity and reducing power of the extracts and found that the results depended on the kind of extracted matrices.

The flavonoid constituents such as amentoflavone, quercitrin, myricitrin, and hinokiflavone have diverse pharmacological properties [134,135]. The paper of Zhuang et al. [67] is devoted to the extraction of myricitrin, quercitrin, amentoflavone, and hinokiflavone from *Platycladi Cacumen*. They found that by applying 12 tested DESs, the yield of myricitrin and quercitrin was higher than that reached using water or methanol as extractants. For the other compounds, the yield was similar. In addition, it was observed that the extraction efficiency is strongly influenced by the viscosity of the used DESs. Saccharides-based DESs showed lower extraction efficiency than that obtained by acid-, amide-, and/or alcohol-based DESs.

7.5. Curcumin

Curcumin is a key active yellow polyphenolic constituent of Curcuma longa; it has various pharmacologic effects including anti-inflammatory, antioxidant, antiproliferative, and antiangiogenic activities [54,136,137]. In traditional medicine (Ayurvedic therapy), it is used to treat stomach disorders, blood cleaning, and skin diseases, as well as disorders of bile production, anorexia, rhinitis, liver functions, and rheumatism or cough [137].

Aydin et al. [86] reported a powerful microextraction method (vortex-assisted DESs emulsification liquid–liquid microextraction) for target compounds curcumin in turmeric drug (food supplement), turmeric powder, and herbal tea. It was shown that this method has the most positive impact at the following conditions: pH (optimal 4), molar ratio (the best results for choline chloride:phenol (1:4)), ratio of DES to tetrahydrofuran (1:1), vortex time (2 min), centrifugation time (5 min at 4500 rpm), and preconcentration factor 12.5. The negative effect of ions in the matrix on efficiency of extraction was confirmed.

7.6. Caffeoylquinic Acids

Caffeoylquinic acids and their derivatives are biologically active dietary polyphenols, playing therapeutic roles such as antioxidant activity, antibacterial, hepatoprotective, cardioprotective, anti-inflammatory, antipyretic, neuroprotective, antiobesity, antiviral, antimicrobial, antihypertension, free radicals scavenger, and a central nervous system stimulator [138,139].

A DES (choline chloride:1,3 butanediol), coupled with the aqueous two-phase system for the negative pressure cavitation extraction, was investigated as a new system for blueberry leaves extraction. Response surface methodology was used to find the best extraction conditions. The target compound (chlorogenic acid) reached the yield of 46.88 mg/g for optimized conditions [82]. In addition, the effects of other techniques such as heat reflux extraction (HRE, 60 °C, 3 h), NPCE (59 °C, 24 min), UAE (60 °C, 60 min, 250 W), and microwave-assisted extraction (MAE, 60 °C, 20 min, 700 W) were compared. The most suitable of the techniques from the viewpoint of chlorogenic acid extraction was NPCE, followed by MAE, HRE, and UAE.

7.7. Isoflavones

Isoflavones such as genistein, daidzein, and biochanin A are widely consumed phytoestrogens. Bajkacz and Adamek [75] investigated the impact of 17 different DESs on the extraction of substances such as genistein, daidzein, genistin, biochanin A, and daidzin from soy products. Stemming from the screening evaluation, choline chloride:citric acid (1:1) was identified as the best of the tested DESs. The influence of different solid to solvent ratios (S/L) and water content was followed, and central composite design on the determination of optimum conditions of extraction was used. The work is unique also due to the fact that the authors compared the extraction efficiency for 10 extraction techniques using several conventional solvents such as methanol, ethanol, different ethanol/methanol mixtures, dimethyl sulphoxide:acetonitrile:water, acetone:hydrochloric acid (5:1 *v/v*). Soxhlet extraction and UAE and MAE methods coupled with HPLC-DAD provided lower yields with higher relative standard deviations in comparison with the DES-UHPLC-UV.

7.8. Rutin

Rutin is the glycoside combining the flavonol quercetin and the disaccharide rutinose. It is a citrus flavonoid found in a wide variety of plants including citrus fruit. It has been explored for a number of pharmacological and nutraceutical effects [140].

Huang et al. [108] investigated the possibilities to apply UAE extraction using 13 DESs (Table 1), and compared the yield of rutin from tartary buckwheat hulls with conventional solvent (80% wt methanol). The ChCl-based DESs with sucrose, sorbitol, and glycerol:glycine, L-histidine:glycerol achieved a lower extraction efficiency than the mentioned conventional solvent. In addition, the authors performed tests on DESs biodegradation and found that all the DESs underwent biodegradability higher than 70% until 28 days.

ChCl with urea, sorbitol, 1,4-butanediol, lactic acid or levulinic acid was used to extract rutin and rosmarinic acid from *Satureja montana* [107]. The amount of rutin obtained was 1.40 to 17.29 mg/g per plant and 0.21 to 7.84 mg of rosmarinic acid/g per plant. Of the solvents, ChCl and lactic acid (1:2) and ChCl:levulinic acid (1:2) were the most suitable for rutin extraction. For rosemary acid, a urea-containing DES proved to be the most suitable. The analysis of the main components showed that increasing the extraction temperature and decreasing the amount of water can increase the extraction of secondary metabolites of the monitored substances.

As documented by screening tests, the extraction of caffeoylmalic acid, psoralic acid-glucoside, rutin, psoralen, and bergapten from *Ficus carica* L. (leaves) was more effective using methanol than eight DESs (Table 1). However, when applying ternary DESs mixtures of glycerol:xylitol:D-(−)-fructose with varying content of individual saccharide components, extraction efficiency exceeded that reached using methanol. It was also documented that, along with extractants, extraction techniques play a significant role in extraction efficiency. Comparing the results obtained by DES-UAE, DES-MAE, methanol-UAE, and methanol-MAE techniques, DES-MAE was identified as the most suitable. The differences in extraction yield were ascribed to differing penetration of the extractant to the matrix [74]. Under optimal conditions (glycerol:xylitol:D-(−)-fructose; extraction temperature 64.46 °C, S/L 1:17.53, and ultrasonic time 24.43 min) the extraction yield of caffeoylmalic acid, psoralic acid-glucoside, rutin, psoralen, and bergapten was 6.482, 16.34, 5.207, 15.22, and 2.475 mg/g, respectively. It is worth mentioning the number of significant figures of the values of the given quantities is the result of optimization by the response surface methodology, not of an experiment.

7.9. Hesperidin

The antioxidant hesperidin, a major flavonoid in orange and lemon, has many pharmacological effects, such as antioxidation, anticancer, antiviral, antibacterial, and anti-inflammatory. The neuroprotective potential of this flavonoid is mediated by the improvement of neural growth factors and endogenous antioxidant defense functions, diminishing neuro-inflammatory and apoptotic pathways [141,142].

Jokic et al. [92] screened the effect of 15 ChCl-based DESs and monitored extraction efficiency from shards of different mandarin varieties (*Okitsu*, *Chahara*, *Kuno*, and *Zorica rana*). The yield of hesperidin ranged from 1.4 to 112 mg/g material. In addition to the screening evaluation, the work focused on determining the optimal conditions for the extraction of hesperidin for these varieties using ChCl and acetamide (1:2). For the *Okitsu* variety, the optimal conditions for hesperidin extraction were 90 min, temperature 68.14 °C, and water content 13.83%; for the variety *Chahara*, it was 45.40 min, 69.70 °C, and a water content of 10.67%, while for the varieties *Kuno* and *Zorica rana*, it was: 88.79 and 54.72 min, 55.02 and 69.66 °C, and 19.73 and 14.86% water, respectively, for extraction of hesperidin with ChCl and acetamide in a molar ratio of 1:2.

7.10. Terpenes

Terpenes are classified according to the number of isoprene units into monoterpenes, sesquiterpenes, diterpenes, and triterpenes. A broad range of the biological activities of plant terpene metabolites are described, including cancer chemopreventive effects, antimicrobial, antifungal, antiviral, antihyperglycemic, anti-inflammatory, and antiparasitic activities [143,144].

7.11. Ginkgolides

Ginkgolides are biologically active terpenic lactones. Su et al. [90] compared the extraction efficiency of 16 DESs (Table 1) and conventional solvents (water, ethanol) for extraction of *Gingko Biloba* leaves (UAE, 70% (w/w) aqueous solution at 100 W and 25 °C for 10 min with a solid to solvent ratio of 1:15). Using a colorimetric method (determination of ginkgolide A), they determined the extraction yield and found that in the case of ChCl:urea and betaine:ethylene glycol, the extraction yield (1.06 mg/g) and (1.15 mg/g), respectively, exceeded that reached using ethanol (1.04 mg/g). Moreover, they followed the influence of water content in the two DESs, and the best extraction was achieved using 40% water content for both DESs. Their paper brings also results from the viewpoint of comparison with other extraction techniques: boiling reflux (60 min, methanol, yield 1.51 mg/g; ethanol, 1.72 mg/g; methanol:water (7:3 v/v), 2.02 mg/g; ethanol:water (7:3 v/v), 2.15 mg/g); betaine:ethyleneglycol + water (6:4, w/w, UAE, 45 °C, 100 W, 20 min, 2.36 mg/g); magnetic stirring (45 °C, 20 min, 2.25 mg/g); and ethanol:water (7:3 v/v, UAE, 45 °C, 100 W, 20 min, 1.84 mg/g). Based on the results obtained, it was concluded that betaine:ethylene glycol represented the most suitable extraction system both in association with UAE and with magnetic stirring. It should be pointed out that the extraction yields are very similar; however, when working at a large scale, the economic benefits may not be negligible.

7.12. Glycyrrhetinic Acid

Glycyrrhetinic acid is a triterpenoid derivative and has different pharmacological properties with possible antiviral, antifungal, antiprotozoal, and antibacterial activities [145]. Sorbitol-based DESs have been used to extract biologically active substances (glycyrrhetinic acid, licuroside) from liquorice root (*Glycyrrhizae*) [91]. Simple maceration was used in this study, and substances such as malic acid, water, and glycerin were used as additional components of DESs.

7.13. Artemisinin

Artemisinin is a sesquiterpene lactone and its derivatives are essential components of antimalarial treatment [146]. Artemisinin is effective also in treating other parasitic diseases, some viral infections, and various neoplasms [147].

Cao et al. [70] realized a screening test of the influence of various methyl trioctyl ammonium chloride-based 13 DESs on the extraction of *Artemisia annua* leaves. The extraction yield ranged from 1 to 1.62 mg/g. Moreover, the impact of molar ratio of 3 two-component (in total, 48 extractants differing in composition) and 15 ternary DESs (in total, 60 extractants) on extraction efficiency was followed (Table 1). The yield ranged from 1.1 to 2.2 mg/g. In particular, the yield of extractive compounds for three different DESs was determined by applying various extraction techniques, namely extraction at

30 or 60 °C by air-bath shaking at 250 rpm, water-bath shaking at 150 rpm, magnetic stirring at 150 rpm, ultrasonication at 200 W, and heating at 60 °C and 0 rpm. The highest yield was reached using UAE at 30 or 60 °C for all tested DESs. Of course, the higher temperature supports penetration of extractant due to lowering its viscosity and, slightly also, its density into the matrices. Subsequently, through an RSM optimization procedure, the influence of such parameters as solid to solvent ratio, ultrasonic power, temperature, particle size, and time of extraction for DES (methyl trioctyl ammonium chloride:1-butanol (1:4)) was evaluated. At optimum conditions (S/L 1:15.5; ultrasonic power 180 W; particle size 80 mesh; temperature 45 °C; time 70 min), the yield reached 7.99 mg/g. It was, thus, confirmed that the UAE method with a selected DES is more efficient than extraction by petroleum ether.

7.14. Polyprenol Acetates

Polyprenol acetates are important lipids with many bioactive and pharmacological activities [148]. Leaves of Gingko biloba were extracted by DESs by Cao et al. [99]. Hydrophobic DESs (15 different DESs) were rated based on polyprenyl acetates extraction. Three of the most effective DESs: methyl trioctyl ammonium chloride:hexyl alcohol; methyl trioctyl ammonium chloride:capryl alcohol; methyl trioctyl ammonium chloride:decyl alcohol were subjected to a more detailed investigation, varying the molar ratio from 1:1 to 1:8. For all selected DESs, the best results were obtained at the 1:5 ratio. Along with the mentioned two-component DESs, ternary systems methyl trioctyl ammonium chloride/capryl alcohol with different second hydrogen bonding donors at different molar ratios (Table 1) were studied. The best DES of them (methyl trioctyl ammonium chloride:capryl alcohol:octyl acid (1:2:3)) was subsequently evaluated from the viewpoint of extraction yield with different extraction methods (25, 60 °C). The gradual decrease in extraction yield for both temperatures was as follows: stirring, UAE, air-bath shaking; water-bath shaking, and heating. Applying RSM, analysis of optimum conditions for the mentioned best DES (84.11 mg/g) was performed and compared to those for other extractants (ethyl acetate (84.17 mg/g); n-hexane (75.48 mg/g), petroleum ether (71.39 mg/g)).

7.15. Proteins

Proteins—large biomolecules consisting of one or more long chains of amino acid residues—play many critical roles in all living organisms. They are an irreplaceable part of food. Their use in medical therapy requires their isolation in pure form [149]. Wahlström et al. [101] extracted Brewer´s spent grain using four eutectic mixtures (sodium formate: urea; potassium acetate: urea; sodium acetate:urea in molar ratios 1:2 or 1:3, and ChCl: urea (1:2)). As a product, proteins composed of amino acids, predominantly serine, arginine, aspartic acid, threonine, alanine, valine, and leucine were isolated. The authors pointed out an advantage of applying a breakthrough technology suitable for the extraction of proteins from various kinds of protein-rich biomass.

8. Factors Limiting the Potential of Deep Eutectic Solvents Utilization and How to Overcome Them

The data in Table 1 and in the previous section documented the pros in the use of DESs for the extraction of value-added substances from phytomass. However, it should be admitted that DESs are not perfect and their use has its limitations. This section deals with the cons of applying DES for extraction purposes and valorization of phytomass.

The use of DESs in the field of biomass pretreatment or extraction of value-added substances has significantly expanded in the 21st century [11]. However, the process of applying faces several limitations from an experimental and commercial point of view.

8.1. Purity

It is natural that the process of developing new types of solvents and their application at a laboratory scale takes place in glass and using analytical grade chemicals. One of the main limits of subsequent commercial application is the possibility of using chemicals with purity lower than analytical grade. Here, however, problems can arise in terms of the stability of the created system. Even though a DES is formed, crystallization may occur due to long-term storage (sometimes, only a few hours). This effect may be exacerbated by impurities that would be present in the starting chemicals and could initiate crystallization. The reason for using chemicals with lower purity is, of course, that their prices are lower than that of pure chemicals. On the other hand, it should be noted that if the DESs application process is commercialized and expanded, the cost of producing DESs will decrease significantly [40]. In this respect, and given the relatively easy and simple preparation, lower costs would help to expand the use of DES as a new way of exploiting the potential of biomass or biowaste. However, it should be noted that the cost of some conventional organic solvents may be lower.

8.2. Viscosity

A significant limiting factor associated with the application of DESs is viscosity. Due to the formation and interaction of hydrogen bonds in the DES structure, the viscosity of DESs is relatively high, being 100–1000 times higher than that of water or conventional organic solvents [150]. On the one hand, viscosity presents the limit for penetration into the substrate, and on the other hand, from the point of view of commercial application, there is a problem in terms of the technological steps associated with the preparation of DESs themselves. This is mainly related to handling, mixing, filling, or transportation. Naturally, there are strategies that can partially eliminate this shortcoming, but the price associated with these measures and the consequent effectiveness of the use of DESs in the required operation with it (the goal of using DESs) play an important role here. The easiest way, although not the cheapest, is to increase the temperature and thus, achieve a decrease in viscosity. Another possibility of a simple solution is to add another reagent to the system, either water or another solvent, that will ensure a decrease in viscosity (e.g., alcohols, [113]). However, whether it is water or another type of solvent, it is necessary to realize that from this point of view, there is a change in the whole system, because the addition of another component also changes the behavior of DES. The addition of water into DESs in the process of their formation causes the incorporation of water molecules into the structure of DESs and their fixation by hydrogen bonds; this water can hardly be later fully removed by, e.g., a rotary evaporator. A small addition of water may result in a decrease in viscosity, temperature lowering, and shorter time needed for DES preparation. Water as another component plays an important role for the formation of hydrogen bond donors and acceptors in the DES structure [151,152]. If DES systems contain water or other organic substances as solvents, it is necessary to take water (for example, organic solvents are automatically taken into account) into account as another component of the DES. Therefore, binary systems need to be characterized as ternary.

8.3. Hygroscopicity

Another challenge or limiting factor is the hygroscopicity of DES. As already mentioned, the addition of water affects the nature of DES in terms of the structure and bonds they form, and water also affects the polarity and ability of DES to extract or solubilize target groups of substances isolated from phytomass. The hygroscopicity of DES must, therefore, not be neglected in the case of ensuring the technological process and its laboratory or commercial use. A more detailed description effect of water vapor from the surrounding air and of these facts is discussed elsewhere [153].

8.4. Long-Term Stability

In order to ensure better handling, mixing, and transport from DES, we have introduced as one of the options an increase in temperature. Regardless of the economic side of things, however, it should be noted that exposure of DES to higher temperatures for a long time can have adverse consequences [11].

8.5. Acid-Base Properties

Acidity or alkalinity are other important factors influencing the applicability of DES. Some DESs have significantly low pH, which significantly limits the choice of materials for their commercial use. Laboratory experiments are usually performed in glass, where this effect can be neglected. When using materials containing different types of metals and their compounds, these can cause an undesirable color change and affect the effect of DESs. The absorption of metallic components from the operating equipment is one of the key problems of the commercial use of DESs. Eliminating this problem may require the use of more expensive materials to transport, mix and apply the DESs themselves, which clearly increases input costs and may potentially discourage potential operators from commercializing the use of DESs. In addition, impurities can destabilize DESs and cause them to crystallize, thereby altering their stability.

8.6. Toxicity

Among the most common issues of research teams, scientists, but also practice and control bodies is the toxicity and recyclability of DES. As for toxicity: at the beginning of the research and application of DESs in 2003, it was very often said that DESs are non-toxic. Over time and the natural evolution of the composition of DESs, this concept has gradually disappeared and currently, DESs are characterized as having low or acceptable toxicity to various biological systems. The toxicity of DESs depends mainly on the toxicity of the starting components, but some DESs may be more toxic than their starting components [153]. The answer to this question about toxicity is a bit unclear. The shortcomings of DESs are gradually emerging, especially in terms of their impact on organisms and the environment; however, the boundaries of the terminology of the impact of DESs and its toxicity are gradually shifting. In general, the toxicity of a substance to organisms and the environment depends on the dose (concentration) and duration of its action. Related to this are the issues of biocompatibility and biodegradability of DESs before applying them to commercial purposes.

8.7. Adsorbable Organic Halides

A question or possibility of other research activities that still arises from published works or relevant project activities is the ability of chlorine-containing DESs to react with a substrate matrix or extractables, leading to adsorbable organic halides. This issue is extremely important in view of the need to limit the use of these halides and even to reject them on the basis of the 12 principles of environmental chemistry in the field of green technologies. However, it should be emphasized that in assessing the possible negative impacts of chlorine compounds, a distinction must be made between "inorganic, ionic" chlorine in the form of Cl^- anions and chlorine bound to a carbon atom in organic compounds. This distinction is important e.g., in waste incineration.

8.8. Recycling

Given the ongoing research and commercial implementation of DESs processes, recyclability issues also need to be answered. Based on the information from the works that dealt with the careful application of DESs, the following conclusion can be drawn. The most common technique in regenerating DESs is to use an anti-solvent to remove (precipitate) the component from the system in operation, and then, evaporate the anti-solvent from the system, and reuse the DESs. Regeneration and reuse aspects are crucial in assessing sustainability and environmental protection, as well as in reducing the costs of the process [11].

9. Future Trends and Concluding Remarks

The excellent properties of DESs, such as sustainability, biodegradability, pharmaceutical acceptable toxicity, negligible volatility, and high extractability of compounds with diverse polarity, highlight their potential as green extractants. Several comparisons of the isolation value-added substances from phytomass performed by DESs and organic solvents have clearly demonstrated, along with ecological advantages, also a higher yield of extracted substances using DESs and thus cost-related benefits. It can be expected that the valorization of phytomass in the future will focus mainly on the extraction of therapeutically important substances, nutrients, and food supplements. The selection and composition of DESs will be optimized so that whole extracts can be used in practice, without prior separation of DESs and extracted value-added substances. It can be assumed that in the field of research of DESs themselves, mixtures with lower viscosity, predetermined polarity, and acid-base properties, capable of specifically extracting targeted value-added substances, will be sought.

Despite the considerable number of phytomass kinds valorized using DESs, there is still a huge amount of primary phytomass itself, waste of its processing, and food-related waste, which have not been studied from the point of view of isolating value-added substances. Expanding resources is a challenge for both laboratory and industrial workers and can bring many surprising and useful results in the future.

Author Contributions: M.J. and J.Š. contributed equally to the conceptualization and design of the work; writing—original draft preparation, M.J. and J.Š. Writing—review and editing; supervision and critical revision of the manuscript, M.J. and J.Š.; and funding acquisition, M.J. All authors have read and agreed to the published version of the manuscript.

Funding: This work was supported by the Slovak Research and Development Agency under the contracts Nos. APVV-15-0052 (50%), and VEGA 1/0403/19 (50%).

Acknowledgments: The authors would like to acknowledge the financial support by the Slovak Research and Development Agency.

Conflicts of Interest: The authors declare no conflict of interest.

Abbreviations

[N(Me)(Oc)$_3$]Cl	methyl trioctyl ammonium chloride
[N(Bu)$_4$]Br	tetrabutylammonium bromide
[N(Pr)$_4$]Br	tetrapropylammonium bromide
ChCl	choline chloride
DES	deep eutectic solvent
dw	dry weight
EAE	enzyme-assisted extraction
GAE	gallic acid equivalents
HBA	hydrogen bond acceptor
HBD	hydrogen bond donor
HDE	hydrodiffusion extraction
HPLC-ESI-TOF-MS	high performance liquid chromatography coupled to electrospray ionization time-of-flight mass spectrometry
HPLC-PDA-ESI-IT/MS	high-performance liquid chromatography coupled to photo diode array detector and electrospray ion-trap mass spectrometry
HPLC-ESI-QTOF-MS	high-performance liquid chromatography coupled to electrospray ionization quadrupole time-of-flight high-resolution mass spectrometry
HPLC-ELSD	high performance liquid chromatography-evaporative light scattering detector method
LC-DAD-MS	high performance liquid chromatographic method coupled with diode-array detection and mass spectrometry

LTTMs	low-transition temperature mixtures
LMMs	low-melting mixtures
MAE	microwave-assisted extraction
NADES	natural deep eutectic solvent
NPC	negative pressure cavitation
PLE	pressurized liquid extraction
RSM	response surface methodology
RtE	rutin equivalents
S/L	solid to solvent ratio
SFE	supercritical fluid extraction
VA-DES-DLLME	vortex assisted deep eutectic solvent dispersive liquid-liquid microextraction
UAE	ultrasound-assisted extraction
UPHLC-Q-TOF-MS	ultra-high performance liquid chromatography-quadrupole time-of-flight mass spectrometry

References

1. Hall, D.O.; House, J.I. Trees and biomass energy: Carbon storage and/or fossil fuel substitution? *Biomass Bioenergy* **1994**, *6*, 11–30. [CrossRef]
2. Bracmort, K. *Biomass: Comparison of Definitions in Legislation*; Congressional Research Service 7-5700, CRS Report R40529; Congressional Research Service: Washington, DC, USA, 2013.
3. FitzPatrick, M.; Champagne, P.; Cunningham, M.F.; Whitney, R.A. A biorefinery processing perspective: Treatment of lignocellulosic materials for the production of value-added products. *Bioresour. Technol.* **2010**, *101*, 8915–8922. [CrossRef] [PubMed]
4. Bar-On, Y.M.; Phillips, R.; Milo, R. The biomass distribution on Earth. *Proc. Natl. Acad. Sci. USA* **2018**, *115*, 6506–6511. [CrossRef] [PubMed]
5. Walden, P. Über die Molekulargrösse und elektrische Leitfähigkeiteiniger geschmolzener Salze. *Bull. Acad. Imper. Sci.* **1914**, *8*, 405–422.
6. Ignatyev, I.A. Cellulose Valorisation in Ionic Liquids. Ph.D. Thesis, Katholieke Universiteit Leuven, Leuven, Belgium, 2011.
7. Xia, Z.; Li, J.; Zhang, J.; Zhang, X.; Zheng, X.; Zhang, J. Processing and Valorization of Cellulose, Lignin and Lignocellulose Using Ionic Liquids. *J. Bioresour. Bioprod.* **2020**, *5*, 79–98. [CrossRef]
8. Kalhor, P.; Ghandi, K. Deep Eutectic Solvents for Pretreatment, Extraction, and Catalysis of Biomass and Food Waste. *Molecules* **2019**, *24*, 4012. [CrossRef]
9. Dai, Y.T.; van Spronsen, J.; Witkamp, G.J.; Verpoorte, R.; Choi, Y.H. Natural deep eutectic solvents as new potential media for green technology. *Anal. Chim. Acta* **2013**, *766*, 61–68. [CrossRef]
10. van Osch, D.J.; Zubeir, L.F.; van den Bruinhorst, A.; Rocha, M.A.; Kroon, M.C. Hydrophobic deep eutectic solvents as water-immiscible extractants. *Green Chem.* **2015**, *17*, 4518–4521. [CrossRef]
11. Jablonský, M.; Šima, J. *Deep Eutectic Solvents in Biomass Valorization*; Spektrum STU: Bratislava, Slovakia, 2019; p. 176. ISBN 978-80-227-4911-4.
12. Liu, P.; Hao, J.W.; Mo, L.P.; Zhang, Z.H. Recent advances in the application of deep eutectic solvents as sustainable media as well as catalysts in organic reactions. *RSC Adv.* **2015**, *5*, 48675–48704. [CrossRef]
13. Manousaki, A.; Jancheva, M.; Grigorakis, S.; Makris, D.P. Extraction of antioxidant phenolics from agri-food waste biomass using a newly designed glycerol-based natural low-transition temperature mixture: A comparison with conventional eco-friendly solvents. *Recycling* **2016**, *1*, 194. [CrossRef]
14. Kottaras, P.; Koulianos, M.; Makris, D.P. Low-Transition Temperature Mixtures (LTTMs) Made of Bioorganic Molecules: Enhanced Extraction of Antioxidant Phenolics from Industrial Cereal Solid Wastes. *Recycling* **2017**, *2*, 3. [CrossRef]
15. Florindo, C.; Lima, F.; Ribeiro, B.D.; Marrucho, I.M. Deep eutectic solvents: Overcoming 21st century challenges. *Curr. Opin. Green Sustain. Chem.* **2019**, *18*, 31–36. [CrossRef]
16. Abbott, A.P.; Capper, G.; Davies, D.L.; Rasheed, R.K.; Tambyrajah, V. Novel solvent properties of choline chloride/urea mixtures. *Chem. Commun.* **2003**, *1*, 70–71. [CrossRef] [PubMed]
17. Smith, E.L.; Abbott, A.P.; Ryder, K.S. Deep eutectic solvents (DESs) and their applications. *Chem. Rev.* **2014**, *114*, 11060–11082. [CrossRef] [PubMed]

18. Abranches, D.O.; Martins, M.A.R.; Silva, L.P.; Schaeffer, N.; Pinho, S.P.; Coutinho, J.A.P. Phenolic hydrogen bond donors in the formation of non-ionic deep eutectic solvents: The quest for type V DES. *Chem. Commun.* **2019**, *55*, 10253–10256. [CrossRef] [PubMed]
19. Usanovich, M. On the "deviations" from Raoult's law due to chemical interaction between the components. *Dokl. Akad. Nauk SSSR* **1958**, *120*, 1304–1306.
20. Lazerges, M.; Rietveld, I.B.; Corvis, Y.; Céolin, R.; Espeau, P. Thermodynamic studies of mixtures for topical anesthesia: Lidocaine–salol binary phase diagram. *Thermochim. Acta* **2010**, *497*, 124–128. [CrossRef]
21. Suriyanarayanan, S.; Olsson, G.; Kathiravan, S.; Ndizeye, N.; Nicholls, I. Non-Ionic Deep Eutectic Liquids: Acetamide–Urea Derived Room Temperature Solvents. *Int. J. Mol. Sci.* **2019**, *20*, 2857. [CrossRef]
22. Nicholls, I.A.; Suriyanarayanan, S. Non-Ionic Deep Eutectic Mixtures for Use as Solvents and Dispersants. International Application No. PCT/SE2019/050161. Patent Number WO20191664442, 22 February 2019. Available online: https://patentscope2.wipo.int/search/en/detail.jsf?docId=WO2019164442 (accessed on 7 August 2020).
23. Skarpalezos, D.; Detsi, A. Deep eutectic solvents as extraction media for valuable flavonoids from natural sources. *Appl. Sci.* **2019**, *9*, 4169. [CrossRef]
24. Häckl, K.; Kunz, W. Some aspects of green solvents. *C. R. Chim.* **2018**, *21*, 572–580. [CrossRef]
25. Liu, Y.; Friesen, J.B.; McAlpine, J.B.; Lankin, D.C.; Chen, S.N.; Pauli, G.F. Natural deep eutectic solvents: Properties, applications, and perspectives. *J. Nat. Prod.* **2018**, *81*, 679–690. [CrossRef] [PubMed]
26. Zhang, Q.; De Oliveira Vigier, K.; Royer, S.; Jérôme, F. Deep eutectic solvents: Syntheses, properties and applications. *Chem. Soc. Rev.* **2012**, *41*, 7108–7146. [CrossRef] [PubMed]
27. Škulcová, A.; Ház, A.; Majova, V.; Sima, J.; Jablonsky, M. Long-term Isothermal Stability of Deep Eutectic Solvents. *BioResources* **2018**, *13*, 7545–7559.
28. Škulcová, A.; Russ, A.; Jablonsky, M.; Šima, J. The pH Behavior of Seventeen Deep Eutectic Solvents. *BioResources* **2018**, *13*, 5042–5051.
29. Jablonsky, M.; Skulcova, A.; Malvis, A.; Sima, J. Extraction of value-added components from food industry based and agro-forest biowastes by deep eutectic solvents. *J. Biotechnol.* **2018**, *282*, 46–66. [CrossRef] [PubMed]
30. Jablonsky, M.; Majova, V.; Ondrigova, K.; Sima, J. Preparation and characterization of physicochemical properties and application of novel ternary deep eutectic solvents. *Cellulose* **2019**, *26*, 3031–3045. [CrossRef]
31. Florindo, C.; Branco, L.C.; Marrucho, I.M. Quest for Green-Solvent Design: From Hydrophilic to Hydrophobic (Deep) Eutectic Solvents. *ChemSusChem* **2019**, *12*, 1549–1559. [CrossRef]
32. Rogošic, M.; Krišto, A.; Kučan, K.Z. Deep eutectic solvents based on betaine and propylene glycol as potential denitrification agents: A liquid-liquid equilibrium study. *Braz. J. Chem. Eng.* **2019**, *36*, 1703–1716. [CrossRef]
33. Chen, J.; Li, Y.; Wang, X.; Liu, W. Application of Deep Eutectic Solvents in Food Analysis: A Review. *Molecules* **2019**, *24*, 4594. [CrossRef]
34. Abbott, A.; Capper, G.; Davies, D.; Rasheed, R.; Shikotra, P. Selective Extraction of Metals from Mixed Oxide Matrixes Using Choline-Based Ionic Liquids. *Inorg. Chem.* **2005**, *44*, 6497–6499. [CrossRef]
35. Triaux, Z.; Petitjean, H.; Marchioni, E.; Boltoeva, M.; Marcic, C. Deep eutectic solvent–based headspace single-drop microextraction for the quantification of terpenes in spices. *Anal. Bioanal. Chem.* **2020**, *412*, 933–948. [CrossRef] [PubMed]
36. Li, X.; Row, K.H. Development of deep eutectic solvents applied in extraction and separation. *J. Sep. Sci.* **2016**, *39*, 3505–3520. [CrossRef] [PubMed]
37. Kua, Y.L.; Gan, S. Natural Deep Eutectic Solvent (NADES) as a Greener Alternative for the Extraction of Hydrophilic (Polar) and Lipophilic (Non-Polar) Phytonutrients. *Key Eng. Mater.* **2019**, *797*, 20–28. [CrossRef]
38. Ruesgas-Ramón, M.; Figueroa-Espinoza, M.C.; Durand, E. Application of deep eutectic solvents (DES) for phenolic compounds extraction: Overview, challenges, and opportunities. *J. Agric. Food Chem.* **2017**, *65*, 3591–3601. [CrossRef]
39. Qin, H.; Hu, X.; Wang, J.; Cheng, H.; Chen, L.; Qi, Z. Overview of acidic deep eutectic solvents on synthesis, properties and applications. *Green Energy Environ.* **2020**, *5*, 8–21. [CrossRef]
40. Jablonský, M.; Škulcová, A.; Šima, J. Use of Deep Eutectic Solvents in Polymer Chemistry—A Review. *Molecules* **2019**, *24*, 3978. [CrossRef]
41. Ruß, C.; König, B. Low melting mixtures in organic synthesis—An alternative to ionic liquids? *Green Chem.* **2012**, *14*, 2969–2982. [CrossRef]

42. Rumble, J.R.; Lide, D.R.; Bruno, T.J. *CRC Handbook of Chemistry and Physics: A Ready-Reference Book of Chemical and Physical Data*; CRC Press: Boca Raton, FL, USA, 2018.
43. Reichardt, C.W.T. *Solvents and Solvent Effects in Organic Chemistry*; John Wiley and Sons: Hoboken, NJ, USA, 2011.
44. Valvi, A.; Dutta, J.; Tiwari, S. Temperature-Dependent Empirical Parameters for Polarity in Choline Chloride Based Deep Eutectic Solvents. *J. Phys. Chem. B* **2017**, *121*, 11356–11366. [CrossRef]
45. Babusca, D.; Benchea, A.C.; Morosanu, A.C.; Dimitriu, D.G.; Dorohoi, D.O. Solvent Empirical Scales and Their Importance for the Study of Intermolecular Interactions. *AIP Conf. Proc.* **2017**, *1796*, 030011.
46. Florindo, C.; McIntosh, A.J.S.; Welton, T.; Branco, L.C.; Marrucho, I.M. A closer look into deep eutectic solvents: Exploring intermolecular interactions using solvatochromic probes. *Phys. Chem. Chem. Phys.* **2018**, *20*, 206–213. [CrossRef]
47. Ramón, D.J.; Guillena, G. *Deep Eutectic Solvents: Synthesis, Propeties, and Applications*; Wiley-VCH: Weinheim, Germany, 2020; ISBN 978-3-527-34518-2.
48. Teles, A.R.R.; Capela, E.V.; Carmo, R.S.; Coutinho, J.A.P.; Silvestre, A.J.D.; Freire, M.G. Solvatochromic parameters of deep eutectic solvents formed by ammonium-based salts and carboxylic acids. *Fluid Phase Equilib.* **2017**, *448*, 15–21. [CrossRef]
49. Yinghuai, Z.; Yuanting, K.T.; Hosmane, N. Applications of ionic liquids in lignin chemistry. In *Ionic Liquids—New Aspects for the Future*; IntechOpen: London, UK, 2013; pp. 315–346.
50. Naser, J.; Mjalli, F.; Jibril, B.; Al-Hatmi, S.; Gano, Z. Potassium Carbonate as a Salt for Deep Eutectic Solvents. *Int. J. Chem. Eng. Appl.* **2013**, *4*, 114–118. [CrossRef]
51. Trajano, H.L.; Wyman, C.E. *Aqueous Pretreatment of Plant Biomass for Biological and Chemical Conversion to Fuels and Chemicals*; John Wiley and Sons: Hoboken, NJ, USA, 2013; pp. 103–128.
52. Atanasov, A.G.; Waltenberger, B.; Pferschy-Wenzig, E.M.; Linder, T.; Wawrosch, C.; Uhrin, P.; Temml, V.; Wang, L.; Schwaiger, S.; Heiss, E.H.; et al. Discovery and resupply of pharmacologically active plant-derived natural products: A review. *Biotechnol. Adv.* **2015**, *33*, 1582–1614. [CrossRef] [PubMed]
53. Dias, D.A.; Urban, S.; Roessner, U. A historical overview of natural products in drug discovery. *Metabolites* **2012**, *2*, 303–336. [CrossRef] [PubMed]
54. Lee, G.; Bae, H. Therapeutic effects of phytochemicals and medicinal herbs on depression. *BioMed Res. Int.* **2017**, *2017*, 6596241. [CrossRef]
55. Jahan, I.; Ahmet, O.N.A.Y. Potentials of plant-based substance to inhabit and probable cure for the COVID-19. *Turk. J. Biol.* **2020**, *44*, 228. [CrossRef] [PubMed]
56. Yonesi, M.; Rezazadeh, A. Plants as a prospective source of natural anti-viral compounds and oral vaccines against COVID-19 coronavirus. *Preprints* **2020**, 2020040321. [CrossRef]
57. David, B.; Wolfender, J.L.; Dias, D.A. The pharmaceutical industry and natural products: Historical status and new trends. *Phytochem. Rev.* **2015**, *14*, 299–315. [CrossRef]
58. Zhang, Z.J. Therapeutic effects of herbal extracts and constituents in animal models of psychiatric disorders. *Life Sci.* **2004**, *75*, 1659–1699. [CrossRef]
59. Wagner, H. Synergy research: Approaching a new generation of phytopharmaceuticals. *Fitoterapia* **2011**, *82*, 34–37. [CrossRef]
60. Havsteen, B.H. The biochemistry and medical significance of the flavonoids. *Pharmacol. Ther.* **2002**, *96*, 67–202. [CrossRef]
61. Wagner, H. Natural products chemistry and phytomedicine in the 21st century: New developments and challenges. *Pure Appl. Chem.* **2005**, *77*, 1–6. [CrossRef]
62. Jablonsky, M.; Nosalova, J.; Sládková, A.; Haz, A.; Kreps, F.; Valka, J.; Miertus, S.; Frecer, V.; Ondrejovic, M.; Sima, J. Valorisation of softwood bark through extraction of utilizable chemicals. A review. *Biotechnol. Adv.* **2017**, *35*, 726–750. [CrossRef] [PubMed]
63. Naoghare, P.K.; Song, J.M. Chip-based high throughput screening of herbal medicines. *Comb. Chem. High Throughout Screen.* **2010**, *13*, 923–931. [CrossRef]
64. Cunha, S.C.; Fernandes, J.O. Extraction Techniques with Deep Eutectic Solvents. *Trends Anal. Chem.* **2018**, *115*, 224–239. [CrossRef]
65. Ivanović, M.; Alañón, M.E.; Arráez-Román, D.; Segura-Carretero, A. Enhanced and green extraction of bioactive compounds from *Lippia citriodora* by tailor-made natural deep eutectic solvents. *Food Res. Int.* **2018**, *111*, 67–76. [CrossRef]

66. Bonacci, S.; Di Gioia, M.L.; Costanzo, P.; Maiuolo, L.; Tallarico, S.; Nardi, M. Natural Deep Eutectic Solvent as Extraction Media for the Main Phenolic Compounds from Olive Oil Processing Wastes. *Antioxidants* 2020, *9*, 513. [CrossRef]
67. Zhuang, B.; Dou, L.L.; Li, P.; Liu, E.H. Deep eutectic solvents as green media for extraction of flavonoid glycosides and aglycones from *Platycladi Cacumen*. *J. Pharm. Biomed.* 2017, *134*, 214–219. [CrossRef]
68. Dedousi, M.; Mamoudaki, V.; Grigorakis, S.; Makris, D.P. Ultrasound-Assisted Extraction of Polyphenolic Antioxidants from Olive (*Olea europaea*) Leaves Using a Novel Glycerol/Sodium-Potassium Tartrate Low-Transition Temperature Mixture (LTTM). *Environments* 2017, *4*, 31. [CrossRef]
69. Fernández, M.L.Á.; Espino, M.; Gomez, F.J.V.; Silva, M.F. Novel approaches mediated by tailor-made green solvents for the extraction of phenolic compounds from agro-food industrial by-products. *Food Chem.* 2018, *239*, 671–678. [CrossRef]
70. Cao, J.; Yang, M.; Cao, F.; Wang, J.; Su, E. Well-Designed Hydrophobic Deep Eutectic Solvents as Green and Efficient Media for the Extraction of Artemisinin from *Artemisia annua* Leaves. *ACS Sustain. Chem. Eng.* 2017, *5*, 3270–3278. [CrossRef]
71. Zhou, P.; Wang, X. Liu, P.; Huang, J.; Wang, C.; Pan, M.; Kuang, Z. Enhanced phenolic compounds extraction from *Morus alba* L. leaves by deep eutectic solvents combined with ultrasonic-assisted extraction. *Ind. Crops Prod.* 2018, *120*, 147–154. [CrossRef]
72. Zhu, S.; Liu, D.; Zhu, X.; Su, A.; Zhang, H. Extraction of Illegal Dyes from Red Chili Peppers with Cholinium-Based Deep Eutectic Solvents. *J. Anal. Methods Chem.* 2017, *2017*, 2753752. [CrossRef] [PubMed]
73. Xiong, Z.; Wang, M.; Guo, H.; Xu, J.; Ye, J.; Zhao, J.; Zhao, L. Ultrasound-assisted deep eutectic solvent as green and efficient media for the extraction of flavonoids from *Radix scutellariae*. *New J. Chem.* 2019, *43*, 644–650. [CrossRef]
74. Wang, T.; Jiao, J.; Gai, Q.Y.; Wang, P.; Guo, N.; Niu, L.L.; Fu, Y.J. Enhanced and green extraction polyphenols and furanocoumarins from Fig (*Ficus carica* L.) leaves using deep eutectic solvents. *J. Pharm. Biomed.* 2017, *145*, 339–345. [CrossRef]
75. Bajkacz, S.; Adamek, J. Evaluation of new natural deep eutectic solvents for the extraction of isoflavones from soy products. *Talanta* 2017, *168*, 329–335. [CrossRef]
76. Fu, N.; Lv, R.; Guo, Z.; Guo, Y.; You, X.; Tang, B.; Row, K.H. Environmentally friendly and non-polluting solvent pretreatment of palm samples for polyphenol analysis using choline chloride deep eutectic solvents. *J. Chromatogr. A* 2017, *1492*, 1–11. [CrossRef]
77. Chanioti, S.; Tzia, C. Extraction of phenolic compounds from olive pomace by using natural deep eutectic solvents and innovative extraction techniques. *Innov. Food Sci. Emerg.* 2018, *48*, 228–239. [CrossRef]
78. Jeong, K.M.; Ko, J.; Zhao, J.; Jin, Y.; Yoo, D.E.; Han, S.Y.; Lee, J. Multi-functioning deep eutectic solvents as extraction and storage media for bioactive natural products that are readily applicable to cosmetic products. *J. Clean. Prod.* 2017, *151*, 87–95. [CrossRef]
79. Yoo, D.E.; Jeong, K.M.; Han, S.Y.; Kim, E.M.; Jin, Y.; Lee, J. Deep eutectic solvent-based valorization of spent coffee grounds. *Food Chem.* 2018, *255*, 357–364. [CrossRef]
80. Duan, L.; Zhang, C.; Zhang, C.; Xue, Z.; Zheng, Y.; Guo, L. Green extraction of phenolic acids from *Artemisia argyi* leaves by Tailor-made ternary deep eutectic solvents. *Molecules* 2019, *24*, 2842. [CrossRef] [PubMed]
81. Vieira, V.; Prieto, M.A.; Barros, L.; Coutinho, J.A.; Ferreira, I.C.; Ferreira, O. Enhanced extraction of phenolic compounds using choline chloride based deep eutectic solvents from *Juglans regia* L. *Ind. Crops Prod.* 2018, *115*, 261–271. [CrossRef]
82. Wang, T.; Xu, W.J.; Wang, S.X.; Kou, P.; Wang, P.; Wang, X.Q.; Fu, Y.J. Integrated and sustainable separation of chlorogenic acid from blueberry leaves by deep eutectic solvents coupled with aqueous two-phase system. *Food Bioprod. Process.* 2017, *105*, 205–214. [CrossRef]
83. Shikov, A.N.; Kosman, V.M.; Flissyuk, E.V.; Smekhova, I.E.; Elameen, A.; Pozharitskaya, O.N. Natural Deep Eutectic Solvents for the Extraction of Phenyletanes and Phenylpropanoids of *Rhodiola rosea* L. *Molecules* 2020, *25*, 1826. [CrossRef]
84. Aryati, W.D.; Nadhira, A.; Febianli, D.; Francisca, F.; Nun'im, A. Natural deep eutectic solvents ultrasound-assisted extraction (NADES-UAE) of trans-cinnamaldehyde and coumarin from cinnamon bark [*Cinnamomum burmannii* (Nees & T. Nees) Blume]. *J. Res. Pharm.* 2020, *24*, 389–398.

85. Sakti, A.S.; Saputri, F.C.; Mun'im, A. Optimization of choline chloride-glycerol based natural deep eutectic solvent for extraction bioactive substances from *Cinnamomum burmannii* barks and *Caesalpinia sappan* heartwoods. *Heliyon* **2019**, *5*, e02915. [CrossRef]
86. Aydin, F.; Yilmaz, E.; Soylak, M. Vortex assisted deep eutectic solvent (DES)-emulsification liquid-liquid microextraction of trace curcumin in food and herbal tea samples. *Food Chem.* **2018**, *243*, 442–447. [CrossRef]
87. Wang, Y.; Peng, B.; Zhao, J.; Wang, M.; Zhao, L. Efficient extraction and determination of prenylflavonol glycosides in *Epimedium pubescens Maxim.* using deep eutectic solvents. *Phytochem. Anal.* **2020**, *31*, 375–383. [CrossRef]
88. Guo, H.; Liu, S.; Li, S.; Feng, Q.; Ma, C.; Zhao, J.; Xiong, Z. Deep eutectic solvent combined with ultrasound-assisted extraction as high efficient extractive media for extraction and quality evaluation of Herba Epimedii. *J. Pharm. Biomed.* **2020**, *185*, 113228. [CrossRef]
89. Wang, X.H.; Wang, J.P. Effective extraction with deep eutectic solvents and enrichment by macroporous adsorption resin of flavonoids from *Carthamus tinctorius* L. *J. Pharm. Biomed.* **2019**, *176*, 112804. [CrossRef]
90. Su, E.; Yang, M.; Cao, J.; Lu, C.; Wang, J.; Cao, F. Deep eutectic solvents as green media for efficient extraction of terpene trilactones from *Ginkgo biloba* leaves. *J. Liq. Chromatogr. Relat. Technol.* **2017**, *40*, 385–391. [CrossRef]
91. Boyko, N.; Zhilyakova, E.; Malyutina, A.; Novikov, O.; Pisarev, D.; Abramovich, R.; Potanina, O.; Lazar, S.; Mizina, P.; Sahaidak-Nikitiuk, R. Studying and Modeling of the Extraction Properties of the Natural Deep Eutectic Solvent and Sorbitol-Based Solvents in Regard to Biologically Active Substances from *Glycyrrhizae Roots*. *Molecules* **2020**, *25*, 1482. [CrossRef] [PubMed]
92. Jokić, S.; Šafranko, S.; Jakovljević, M.; Cikoš, A.M.; Kajić, N.; Kolarević, F.; Babić, J.; Molnar, M. Sustainable Green Procedure for Extraction of Hesperidin from Selected Croatian Mandarin Peels. *Processes* **2019**, *7*, 469. [CrossRef]
93. Faraji, M. Novel hydrophobic deep eutectic solvent for vortex assisted dispersive liquid-liquid micro-extraction of two auxins in water and fruit juice samples and determination by high performance liquid chromatography. *Microchem. J.* **2019**, *150*, 104130. [CrossRef]
94. Li, X.; Row, K.H. Application of deep eutectic solvents in hybrid molecularly imprinted polymers and mesoporous siliceous material for solid-phase extraction of levofloxacin from green bean extract. *Anal. Sci.* **2017**, *33*, 611–617. [CrossRef]
95. Yiin, C.L.; Quitain, A.T.; Yusup, S.; Uemura, Y.; Sasaki, M.; Kida, T. Choline chloride (ChCl) and monosodium glutamate (MSG)-based green solvents from optimized cactus malic acid for biomass delignification. *Bioresour. Technol.* **2017**, *244*, 941–948. [CrossRef]
96. Alishlah, T.; Mun'im, A.; Jufri, M. Optimization of Urea-glycerin Based NADES-UAE for Oxyresveratrol Extraction from *Morus alba* Roots for Preparation of Skin Whitening Lotion. *J. Young Pharm.* **2019**, *11*, 151–156. [CrossRef]
97. Liew, S.Q.; Ngoh, G.C.; Yusoff, R.; Teoh, W.H. Acid and Deep Eutectic Solvent (DES) extraction of pectin from pomelo (Citrus grandis (L.) Osbeck) peels. *Biocatal. Agric. Biotechnol.* **2018**, *13*, 1–11. [CrossRef]
98. Obluchinskaya, E.D.; Daurtseva, A.V.; Pozharitskaya, O.N.; Flisyuk, E.V.; Shikov, A.N. Natural Deep Eutectic Solvents as alternatives for extracting phlorotannins from brown algae. *Pharm. Chem. J.* **2019**, *53*, 243–247. [CrossRef]
99. Cao, J.; Yang, M.; Cao, F.; Wang, J.; Su, E. Tailor-made hydrophobic deep eutectic solvents for cleaner extraction of polyprenyl acetates from *Ginkgo biloba* leaves. *J. Clean. Prod.* **2017**, *152*, 399–405. [CrossRef]
100. Cao, J.; Chen, L.; Li, M.; Cao, F.; Zhao, L.; Su, E. Efficient extraction of proanthocyanidin from Ginkgo biloba leaves employing rationally designed deep eutectic solvent-water mixture and evaluation of the antioxidant activity. *J. Pharm. Biomed.* **2018**, *158*, 317–326. [CrossRef] [PubMed]
101. Wahlström, R.; Rommi, K.; Willberg-Keyriläinen, P.; Ercili-Cura, D.; Holopainen-Mantila, U.; Hiltunen, J.; Mäkinen, O.; Nygren, H.; Mikkelson, A.; Kuutti, L. High Yield Protein Extraction from Brewer's Spent Grain with Novel Carboxylate Salt-Urea Aqueous Deep Eutectic Solvents. *ChemistrySelect* **2017**, *2*, 9355–9363. [CrossRef]
102. Lin, Z.; Jiao, G.; Zhang, J.; Celli, G.B.; Brooks, M.S.L. Optimization of protein extraction from bamboo shoots and processing wastes using deep eutectic solvents in a biorefinery approach. *Biomass Convers. Biorefin.* **2020**, 1–12. [CrossRef]
103. Wang, X.; Li, G.; Ho Row, K. Extraction and Determination of Quercetin from Ginkgo biloba by DESs-Based Polymer Monolithic Cartridge. *J. Chromatogr. Sci.* **2017**, *55*, 866–871. [CrossRef]

104. Cui, Q.; Liu, J.Z.; Wang, L.T.; Kang, Y.F.; Meng, Y.; Jiao, J.; Fu, Y.J. Sustainable deep eutectic solvents preparation and their efficiency in extraction and enrichment of main bioactive flavonoids from sea buckthorn leaves. *J. Clean. Prod.* **2018**, *184*, 826–835. [CrossRef]
105. Meng, Z.; Zhao, J.; Duan, H.; Guan, Y.; Zhao, L. Green and efficient extraction of four bioactive flavonoids from Pollen Typhae by ultrasound-assisted deep eutectic solvents extraction. *J. Pharm. Biomed.* **2018**, *161*, 246–253. [CrossRef]
106. Tian, H.; Wang, J.; Li, Y.; Bi, W.; Chen, D.D.Y. Recovery of natural products from deep eutectic solvents by mimicking denaturation. *ACS Sustain. Chem. Eng.* **2019**, *7*, 9976–9983. [CrossRef]
107. Jakovljević, M.; Vladić, J.; Vidović, S.; Pastor, K.; Jokić, S.; Molnar, M.; Jerković, I. Application of Deep Eutectic Solvents for the Extraction of Rutin and Rosmarinic Acid from Satureja montana L. and Evaluation of the Extracts Antiradical Activity. *Plants* **2020**, *9*, 153. [CrossRef]
108. Huang, Y.; Feng, F.; Jiang, J.; Qiao, Y.; Wu, T.; Voglmeir, J.; Chen, Z.G. Green and efficient extraction of rutin from tartary buckwheat hull by using natural deep eutectic solvents. *Food Chem.* **2017**, *221*, 1400–1405. [CrossRef]
109. Karageorgou, I.; Grigorakis, S.; Lalas, S.; Makris, D.P. Enhanced extraction of antioxidant polyphenols from *Moringa oleifera Lam.* leaves using a biomolecule-based low-transition temperature mixture. *Eur. Food Res. Technol.* **2017**, *243*, 1839–1848. [CrossRef]
110. Kurtulbaş, E.; Pekel, A.G.; Bilgin, M.; Makris, D.P.; Şahin, S. Citric acid-based deep eutectic solvent for the anthocyanin recovery from Hibiscus sabdariffa through microwave-assisted extraction. *Biomass Convers. Biorefin.* **2020**, 1–10. [CrossRef]
111. Lakka, A.; Grigorakis, S.; Kaltsa, O.; Karageorgou, I.; Batra, G.; Bozinou, E.; Lalas, S.; Makris, D.P. The effect of ultrasonication pretreatment on the production of polyphenol-enriched extracts from *Moringa oleifera* L. (drumstick tree) using a novel bio-based deep eutectic solvent. *Appl. Sci.* **2020**, *10*, 220. [CrossRef]
112. Pavić, V.; Flačer, D.; Jakovljević, M.; Molnar, M.; Jokić, S. Assessment of total phenolic content, in vitro antioxidant and antibacterial activity of Ruta graveolens L. extracts obtained by choline chloride based natural deep eutectic solvents. *Plants* **2019**, *8*, 69. [CrossRef] [PubMed]
113. Jablonsky, M.; Majova, V.; Strizincova, P.; Sima, J.; Jablonsky, J. Investigation of Total Phenolic Content and Antioxidant Activities of Spruce Bark Extracts Isolated by Deep Eutectic Solvents. *Crystals* **2020**, *10*, 402. [CrossRef]
114. Torres-Vega, J.; Gómez-Alonso, S.; Pérez-Navarro, J.; Pastene-Navarrete, E. Green Extraction of Alkaloids and Polyphenols from *Peumus boldus* Leaves with Natural Deep Eutectic Solvents and Profiling by HPLC-PDA-IT-MS/MS and HPLC-QTOF-MS/MS. *Plants* **2020**, *9*, 242. [CrossRef]
115. Georgantzi, C.; Lichou, A.E.; Paterakis, N.; Makris, D.P. Combination of Lactic Acid-Based Deep Eutectic Solvents (DES) with β-Cyclodextrin: Performance Screening Using Ultrasound-Assisted Extraction of Polyphenols from Selected Native Greek Medicinal Plants. *Agronomy* **2017**, *7*, 54. [CrossRef]
116. Kaltsa, O.; Lakka, A.; Grigorakis, S.; Karageorgou, I.; Batra, G.; Bozinou, E.; Lalas, S.; Makris, D.P. A green extraction process for polyphenols from elderberry (Sambucus nigra) flowers using deep eutectic solvent and ultrasound-assisted pretreatment. *Molecules* **2020**, *25*, 921. [CrossRef]
117. González, C.G.; Mustafa, N.R.; Wilson, E.G.; Verpoorte, R.; Choi, Y.H. Application of natural deep eutectic solvents for the "green" extraction of vanillin from vanilla pods. *Flavour Fragr. J.* **2017**, *33*, 91–96. [CrossRef]
118. Mbous, Y.P.; Hayyan, M.; Hayyan, A.; Wong, W.F.; Hashim, M.A.; Looi, C.Y. Applications of deep eutectic solvents in biotechnology and bioengineering-Promises and challenges. *Biotechnol. Adv.* **2017**, *35*, 105–134. [CrossRef]
119. Perna, F.M.; Vitale, P.; Capriati, V. Deep eutectic solvents and their applications as green solvents. *Curr. Opin. Green Sustain. Chem.* **2020**, *21*, 27–33. [CrossRef]
120. Jablonsky, M.; Majova, V.; Sima, J.; Hrobonova, K.; Lomenova, A. Involvement of Deep Eutectic Solvents in Extraction by Molecularly Imprinted Polymers—A Minireview. *Crystals* **2020**, *10*, 217. [CrossRef]
121. Cory, H.; Passarelli, S.; Szeto, J.; Tamez, M.; Mattei, J. The Role of Polyphenols in Human Health and Food Systems: A Mini-Review. *Front. Nutr.* **2018**, *5*, 87. [CrossRef] [PubMed]
122. Lee, O.H.; Lee, B.Y.; Lee, J.; Lee, H.B.; Son, J.Y.; Park, C.S.; Shetty, K.; Kim, Y.C. Assessment of phenolics-enriched extract and fractions of olive leaves and their antioxidant activities. *Bioresour. Technol.* **2009**, *100*, 6107–6113. [CrossRef] [PubMed]

123. Philippi, K.; Tsamandouras, N.; Grigorakis, S.; Makris, D.P. Ultrasound-assisted green extraction of eggplant peel (*Solanum melongena*) polyphenols using aqueous mixtures of glycerol and ethanol: Optimisation and kinetics. *Environ. Proc.* **2016**, *3*, 369–386. [CrossRef]
124. Eom, S.H.; Kim, Y.M.; Kim, S.K. Antimicrobial effect of phlorotannins from marine brown algae. *Food Chem. Toxicol.* **2012**, *50*, 3251–3255. [CrossRef] [PubMed]
125. Kawamura-Konishi, Y.; Watanabe, N.; Saito, M.; Nakajima, N.; Sakaki, T.; Katayama, T.; Enomoto, T. Isolation of a new phlorotannin, a potent inhibitor of carbohydrate-hydrolyzing enzymes, from the brown alga *Sargassum patens*. *J. Agric. Food Chem.* **2012**, *60*, 5565–5570. [CrossRef]
126. Lopes, G.; Pinto, E.; Andrade, P.B.; Valentao, P. Antifungal activity of phlorotannins against dermatophytes and yeasts: Approaches to the mechanism of action and influence on Candida albicans virulence factor. *PLoS ONE* **2013**, *8*, e72203. [CrossRef]
127. Panche, A.N.; Diwan, A.D.; Chandra, R.S. Flavonoids: An overview. *J. Nutr. Sci.* **2016**, *5*, e47. [CrossRef]
128. Rashid, M.; Fareed, M.; Rashid, H.; Aziz, H.; Ehsan, N.; Khalid, S.; Ghaffar, I.; Ali, R.; Gul, A.; Hakeem, K. Flavonoids and Their Biological Secrets: Phytochemistry and Molecular Aspects. *Plant Hum. Health* **2019**, *2*, 579–605.
129. Li, Y.; Yao, J.; Han, C.; Yang, J.; Chaudhry, M.T.; Wang, S.; Liu, H.; Yin, Y. Quercetin, inflammation and immunity. *Nutrients* **2016**, *8*, 167. [CrossRef]
130. Mlcek, J.; Jurikova, T.; Skrovankova, S.; Sochor, J. Quercetin and its anti-allergic immune response. *Molecules* **2016**, *21*, 623. [CrossRef] [PubMed]
131. Shindel, A.W.; Chen, Z.C.; Lin, G.; Fandel, T.M.; Huang, Y.C.; Banie, L.; Breyer, B.N.; Garcia, M.M.; Lin, C.S.; Lue, T.F. Erectogenic and neurotrophic effects of icariin, a purified extract of horny goat weed (*Epimedium* spp.) in vitro and in vivo. *J. Sex. Med.* **2010**, *7*, 1518–1528. [CrossRef] [PubMed]
132. Jiang, J.; Zhao, B.J.; Song, J.; Jia, X.B. Pharmacology and clinical application of plants in *Epimedium* L. *Chin. Herb. Med.* **2016**, *8*, 12–23. [CrossRef]
133. Jadeja, R.N.; Devkar, R.V. Polyphenols in chronic diseases and their mechanisms of action. *Polyphen. Hum. Health Dis.* **2014**, *1*, 615–623.
134. Meotti, F.C.; Senthilmohan, R.; Harwood, D.T.; Missau, F.C.; Pizzolatti, M.G.; Kettle, A.J. Myricitrin as a substrate and inhibitor of myeloperoxidase: Implications for the pharmacological effects of flavonoids. *Free Radic. Biol. Med.* **2008**, *44*, 109–120. [CrossRef]
135. Shan, C.X.; Guo, S.C.; Yu, S.; Shan, M.Q.; Li, S.F.Y.; Chai, C.; Cui, X.B.; Zhang, L.; Ding, A.W.; Wu, Q.N. Simultaneous Determination of Quercitrin, Afzelin, Amentoflavone, Hinokiflavone in Rat Plasma by UFLC–MS-MS and Its Application to the Pharmacokinetics of *Platycladus orientalis* Leaves Extract. *J. Chromatogr. Sci.* **2018**, *56*, 895–902. [CrossRef]
136. Anand, P.; Kunnumakkara, A.B.; Newman, R.A.; Aggarwal, B.B. Bioavailability of curcumin: Problems and promises. *Mol. Pharm.* **2007**, *4*, 807–818. [CrossRef]
137. Yeung, A.W.K.; Horbańczuk, M.; Tzvetkov, N.T.; Mocan, A.; Carradori, S.; Maggi, F.; Marchewka, J.; Sut, S.; Dall'Acqua, S.; Gan, R.Y.; et al. Curcumin: Total-scale analysis of the scientific literature. *Molecules* **2019**, *24*, 1393. [CrossRef]
138. Naveed, M.; Hejazi, V.; Abbas, M.; Kamboh, A.A.; Khan, G.J.; Shumzaid, M.; Ahmad, F.; Babazadeh, D.; FangFang, X.; Modarresi-Ghazani, F.; et al. Chlorogenic acid (CGA): A pharmacological review and call for further research. *Biomed. Pharmacother.* **2018**, *97*, 67–74. [CrossRef]
139. Tamayose, C.I.; Santos, E.A.; Roque, N.; Costa-Lotufo, L.V.; Ferreira, M.J.P. Caffeoylquinic acids: Separation method, antiradical properties and cytotoxicity. *Chem. Biodivers.* **2019**, *16*, e1900093. [CrossRef]
140. Ganeshpurkar, A.; Saluja, A.K. The pharmacological potential of rutin. *Saudi Pharm. J.* **2017**, *25*, 149–164. [CrossRef] [PubMed]
141. Hajialyani, M.; Hosein Farzaei, M.; Echeverría, J.; Nabavi, S.M.; Uriarte, E.; Sobarzo-Sánchez, E. Hesperidin as a neuroprotective agent: A review of animal and clinical evidence. *Molecules* **2019**, *24*, 648. [CrossRef] [PubMed]
142. Wilmsen, P.K.; Spada, D.S.; Salvador, M. Antioxidant activity of the flavonoid hesperidin in chemical and biological systems. *J. Agric. Food Chem.* **2005**, *53*, 4757–4761. [CrossRef] [PubMed]
143. Paduch, R.; Kandefer-Szerszeń, M.; Trytek, M.; Fiedurek, J. Terpenes: Substances useful in human healthcare. *Arch. Immunol. Ther. Exp.* **2007**, *55*, 315. [CrossRef]

144. Gonzalez-Burgos, E.; Gómez-Serranillos, M.P. Terpene compounds in nature: A review of their potential antioxidant activity. *Curr. Med. Chem.* **2012**, *19*, 5319–5341. [CrossRef]
145. Badam, L. In vitro antiviral activity of indigenous glycyrrhizin, licorice and glycyrrhizic acid (Sigma) on Japanese encephalitis virus. *J. Commun. Dis.* **1997**, *29*, 91–99.
146. White, N.J. Qinghaosu (artemisinin): The price of success. *Science* **2008**, *320*, 330–334. [CrossRef]
147. Weathers, P.J.; Arsenault, P.R.; Covello, P.S.; McMickle, A.; Teoh, K.H.; Reed, D.W. Artemisinin production in Artemisia annua: Studies in planta and results of a novel delivery method for treating malaria and other neglected diseases *Phytochem. Rev.* **2011**, *10*, 173–183. [CrossRef]
148. Tao, R.; Wang, C.Z.; Ye, J.Z.; Zhou, H.; Chen, H.X.; Zhang, C.W. Antibacterial, cytotoxic and genotoxic activity of nitrogenated and haloid derivatives of C 50–C 60 and C 70–C 120 polyprenol homologs. *Lipids Health Dis.* **2016**, *15*, 175. [CrossRef]
149. Stephenson, F.H. *Proteins, in Calculations for Molecular Biology and Biotechnology*, 3rd ed.; Elsevier: Amsterdam, The Netherlands, 2016; pp. 375–429.
150. Tang, B.; Row, K.H. Recent developments in deep eutectic solvents in chemical sciences. *Chem. Mon.* **2013**, *144*, 1427–1454. [CrossRef]
151. Francisco, M.; Bruinhorst, A.V.D.; Kroon, M.C. Low-Transition-Temperature Mixtures (LTTMs): A New Generation of Designer Solvents. *Angew. Chem. Int. Ed.* **2013**, *52*, 3074–3085. [CrossRef] [PubMed]
152. Smith, P.; Arroyo, C.B.; Hernandez, F.L.; Goeltz, J.C. Ternary Deep Eutectic Solvent Behavior of Water and Urea? Choline Chloride Mixtures. *J. Phys. Chem. B* **2019**, *123*, 5302–5306. [CrossRef] [PubMed]
153. Emami, S.; Shayanfar, A. Deep eutectic solvents for pharmaceutical formulation and drug delivery applications. *Pharm. Dev. Technol.* **2020**, 1–18. [CrossRef] [PubMed]

© 2020 by the authors. Licensee MDPI, Basel, Switzerland. This article is an open access article distributed under the terms and conditions of the Creative Commons Attribution (CC BY) license (http://creativecommons.org/licenses/by/4.0/).

Review

Processing of Functional Composite Resins Using Deep Eutectic Solvent

Jing Xue, Jing Wang *, Daoshuo Feng, Haofei Huang and Ming Wang

School of Chemistry and Chemical Engineering, Shandong University of Technology, 266 Xincun Road, Zibo 255000, China; xuejingsdut@163.com (J.X.); fengdaoshuosdut@163.com (D.F.); 1982hhf@163.com (H.H.); wangmingmw@sdut.edu.cn (M.W.)
* Correspondence: wjing@sdut.edu.cn

Received: 24 August 2020; Accepted: 23 September 2020; Published: 24 September 2020

Abstract: Deep eutectic solvents (DESs)—a promising class of alternatives to conventional ionic liquids (ILs) that have freezing points lower than the individual components—are typically formed from two or more components through hydrogen bond interactions. Due to the remarkable advantages of biocompatibility, economical feasibility and environmental hospitality, DESs show great potentials for green production and manufacturing. In terms of the processing of functional composite resins, DESs have been applied for property modifications, recyclability enhancement and functionality endowment. In this review, the applications of DESs in the processing of multiple functional composite resins such as epoxy, phenolic, acrylic, polyester and imprinted resins, are covered. Functional composite resins processed with DESs have attracted much attention of researchers in both academic and industrial communities. The tailored properties of DESs for the design of functional composite resins—as well as the effects of hydrogen bond on the current polymeric systems—are highlighted. In addition to the review of current works, the future perspectives of applying DESs in the processing of functional composite resins are also presented.

Keywords: deep eutectic solvent; composite resins; hydrogen bond

1. Introduction

Composite resins play a significant role in industrial and domestic applications because of their advantages of lightness, plasticity and economic feasibility. The solvents, which plays multiples roles such as diluent, monomer and viscosity modifier, is indispensable in processing of traditional composite resins. However, the usage of volatile organic solvents (VOC) in the processing of traditional composite resins induces serious environmental problems which limit the industrial development of this type of macromolecular material [1,2]. Modifications of traditional composite resins have become the focus of research in recent years. Modified composite resins have brought great convenience to production and life [3].

In the past few decades, research is increasingly focusing on green processing of resin composites in order to deal with the hazards of VOCs to make them more environmentally friendly. Abbott et al. [4] firstly introduced deep eutectic solvent (DES) by mixing choline chloride (ChCl) with urea which has advantages of easy preparation, low cost, non-toxicity and biodegradability. DESs have been regarded as a new generation of ionic liquids (ILs) with a variety of applications in polymer processing, biomedicine and nanotechnology [5]. DESs are usually composed of quaternary ammonium salts and metal salts or hydrogen-bond donors (HBDs), which could be prepared by simply mixing [6]. Before the practice of DESs, ILs took the domination of green solvent for various applications. Tang et al. [7] studied a hydrophobic IL-modified thermo-responsive molecularly imprinted monolith and N-isopropylacrylamide as a thermo-responsive monomer for selective recognition and separation of tanshinones. Due to the porous and low-pressure nature of monolith, the separation of the five tanshinones was achieved via the thermo-responsive hydrophilicity/hydrophobicity transformation in water. Tang et al. [8] also synthesized CO_2-induced switchable ILs with reversible hydrophobic/hydrophilic conversion, and they applied the novel ILs for lipid extraction and separation from wet microalgae by bubbling CO_2. Because of the physiochemical similarity between DESs and ILs, DESs started taking place of ILs. Similar to ILs, due to their solvent properties, DESs can also dissolve CO_2, metal oxides and versatile organic species. The solubility strongly depends on the pressure and temperature [9]. Moreover, DESs have also been applied to the field of catalysis, such as base-catalyzed reactions, acid-catalyzed reactions and transition-metal-catalyzed reactions because of their designable chemical structures and excellent solvent properties [10]. The formation of eutectic mixtures has also provided a new extraction method by modifying the chemical structures of HBDs where DES is formed in situ and extracted, and this method used DESs as extractants for separations of liquid, solid and gaseous phases [11,12]. Lou et al. [13] investigated the extraction of lignin nanoparticles from herbaceous biomass (wheat straw) with ChCl–lactic acid DES. It was found that DES could extract high purity lignin (up to 94.8%) and the water content in biomass affected the hydrogen bond interaction between lignin and DES, which influenced the lignin extraction yield. Tang et al. [14] synthesized novel DESs modified molecularly imprinted polymers (DESs–MIPs) using acrylamide as function monomer, alcohol-based DESs as auxiliary function monomer and chloramphenicol (CAP) as the template. The adsorption results that the ChCl/ethylene glycol DESs-based MIPs had stable interactions with CAP. In addition to the applications in catalysis and separations, DESs have also played an important role in polymer synthesis and processing [15,16], which provided hints and inspirations for the green processing of functional composite resins. Despite of the fact that a relatively smaller number of publications about composite resin processing with DESs are available currently [17,18], it is of great significance to both academia and industry for the green production of composite resins by encompassing research about processing of various composite resins with DESs. This paper covers the main categories of composite resins including epoxy resin, phenolic resin, acrylics, polyester resin and imprinted resin. It is in a trend that DES is playing an all-in-one role in the green processing of composite resins and is providing a promising future on the development of green synthesis and production, which is illustrated in the end as the perspective of this review.

2. Deep Eutectic Solvent

DESs could be defined as a general formula $R_1R_2R_3R_4N^+X^-Y^-$ [19], and they can be divided into three types. When it was discovered that metal salts coupled with alcohols and amides can also form DESs (MCl_x + RZ; M=Al, Zn; Z=$CONH_2$, OH), a fourth type of DES was added [20]. Among the four types of DES, Type III DES (Figure 1a) is the most attractive to researchers, which is typically formed through hydrogen bond, where the charge delocalization occurring through hydrogen bonds between the halide anions and the HBDs leads to the decrease in the freezing point of the mixture (Figure 1b) [17,21]. Taking DES of ChCl/urea as an example, through molecular dynamics simulation and spatial distribution function analysis, some researchers have found that after formation of DES, increases were observed on ions disorder and charge density, while decreases were observed on lattice energy and freezing point [22].

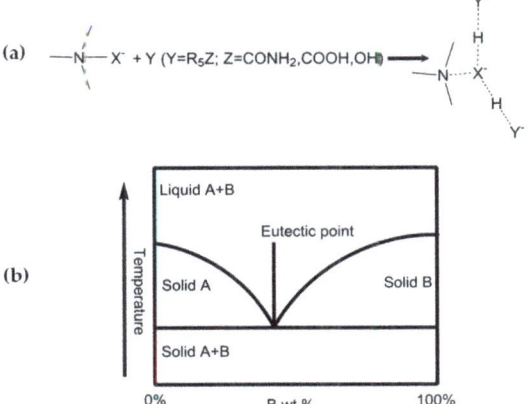

Figure 1. (a) Illustration of Type III deep eutectic solvent with hydrogen bond formed; (b) principle behind Type III deep eutectic solvents.

Structural characterization of DES is usually carried out by using Fourier transform infrared spectrometry (FTIR), nuclear magnetic resonance spectrum (NMR), dynamic mechanical analysis (DMA) and differential scanning calorimetry (DSC). Wang et al. [23] prepared an antibacterial DES based on benzalkonium chloride (BC) and acrylic acid (AA), and they determined the formation of DES by FTIR and NMR. FTIR result showed that the hydrogen bond mainly effected the position of C=O stretch between 1700–1725 cm^{-1} in AA. Single component of AA showed peaks at 1700 cm^{-1} and 1725 cm^{-1}, but the peak at 1700 cm^{-1} diminished and at the shoulder of the peak at 1725 cm^{-1} for AA-derived eutectic mixtures. Similarly, the chemical shift in the NMR of -COOH proton was at δ = 11.38 for AA, and the chemical shift moved to a lower value (δ = 10.58) after mixing with BC. The formation of hydrogen bond for BC–AA had been evidenced by FTIR and NMR. Gautam et al. [24] found that the vibrational peaks in the FTIR spectra of DES formed with ChCl and carboxylic acid shifted to a longer wavelength compared to its individual component, which is a sign of hydrogen bond formation. They further investigated the hydrogen bonds location and length through hydrogen bond sites and confirmed the existence of hydrogen bonds by quantum theory of atoms and the reduced density gradient. Troter et al. [25] reported the effect of temperature on dynamic viscosity of DESs of the binary ChCl-based DESs by DMA at a temperature range of 293.15–363.15 K. the results showed that the viscosity–temperature curve of the DES with ChCl followed the Arrhenius equation, and it was demonstrated that the viscosity of DES was affected by the formation of hydrogen bond. Aroso et al. [26] studied the thermal and rheological properties of DESs, and these DES systems had Newtonian behavior as well as viscosity decrease with temperature and water content. DSC characterization confirmed that for water content at 1:1:1 molar ratio, the mixture retained its single-phase behavior. Tomai et al. [27] prepared a low transition temperature mixture (LTTM) by mixing ChCl and acetylsalicylic acid in a molar ratio 1:2. DSC result showed that only a Tg at −37 °C was observed and the sample did not undergo a phase transition, crystallization or melting, and therefore it was suggested to be defined as a LTTM instead of DES. Francisco et al. [28] found that hydrogen bonding can be evidenced by the shifts in the FTIR and a shift in the resonance signal can also be noticed to lower field in NMR. Choi et al. [29] observed intermolecular interaction between the sucrose and the malic acid mixture by H–H-nuclear, implying that molecules of these compounds in the liquid were aggregated into larger structures. Stefanovic et al. [30] reported that the trends in the solvation structure of poly ethylene glycol (PEO) is closely related to the density of hydrogen bond network within DES and the extent of disturbance induced by the polymer solute. It was indicted that the incorporation of polymers as solute into DES would destruct the complex solvation environment leading to a transformation in polymer structures with relatively static spiral shape. The interactions between the hydrogen bonds within DES and the polymers as solute in DES-based solutions is remaining an interesting topic when using DES to process composite resins.

When DES is employed to modify the composite resins, proper chosen chemical structures of individual components within DES could significantly affect the properties of the original composite resins which may include the hydrophobicity/hydrophilicity, viscosity and conductivity. Tiecco et al. [31] have demonstrated that the hydrophobicity/hydrophilicity of DESs corresponded to the chosen HBDs instead of hydrogen-bond acceptors (HBAs). For example, even when the DES was prepared with a highly water-soluble HBAs, it would still be easily separated from water if the chosen HBD is highly hydrophobic. Tang et al. [32] applied fatty acid/alcohol-based hydrophobic DESs in the extraction of levofloxacin and ciprofloxacin in water by liquid–liquid microextraction. The mechanism of extraction was that the targets can be transferred from the water phase to the DES phase by hydrogen bonding. the extraction efficiency of alcohol-based hydrophobic DESs was higher than that of the fatty acid-based DESs, and because that the long carbon chain structure of fatty acid-based DES increased the distance between the hydroxyl group and HBA, which weakens the hydrogen bond strength. Tang et al. [33] further constructed a polarity controlled biphasic extraction system by combining a hydrophilic DES phase (hexafluoroisopropanol–choline chloride) and a hydrophobic DES phase (menthol–tricaprylylmethylammonium chloride), and a DES-based biphasic system was used to extract and separate high/low polarity compounds. Therefore, the polarity control of DES-based biphasic system can be used to regulate hydrophobicity/hydrophilicity properties of extraction system. Tang et al. [34] also employed a choline salt–aniline DES to remove aniline from organic waste liquid, and the mechanism of extraction was to remove aniline by forming a choline salt–aniline DES and choline-based DESs were insoluble in the organic solvent due to the differences in polarity. The result showed that a choline salt–aniline DES had higher extraction capacity (>95%) than traditional extraction agents. From this point, it would be made possible to adjust the hydrophobicity/hydrophilicity of composite resins through the incorporation of DES. Research on viscosity and conductivity of DESs is also inevitable. Sas et al. [35] measured the viscosity of DES formed with ChCl and levulinic acid, and it was found that the viscosity of DES obeyed the Arrhenius equation. Moreover, the conductivity of DES was related to its viscosity. Abbott et al. [36] further summarized the hole theory claiming that the viscosity and conductivity of DESs were closely related to the dimension of species and the free volume and increased size of free volume was correlated to decreased surface tension. Hole theory is helpful for the design of DESs with favorable viscosity and conductivity, which set the basis for the processing of composite resins with DESs on certain circumstances.

Knowledge of thermal stability of DESs is important for applying them to process composite resins at high temperatures. Chen et al. [37] used thermal gravimetric analysis (TGA) to study the thermal stability of various DESs. By comparing the onset decomposition temperatures of DESs, the reasons of decomposition and weight loss were considered to be due to the destruction of hydrogen bond after being heated. The analysis of the thermal decomposition behavior of DESs provided a basis for preparing DESs with adjustable thermal stability. In order to use DESs for a variety of industrial applications, Ghedi et al. [38] investigated the thermal stability of DESs in a certain temperature range, and they found that the thermal stability of DESs increased with the increase of alkyl chain length of the HBD. They further used FTIR to investigate the vibration modes of hydrogen bond and analyzed the intaeractions of functional groups on HBDs and HBAs. Troter et al. [25] compared the difference of the physical and thermodynamic properties of a series of binary DESs at the temperature range of 293.15–363.15 K. With the increase of temperature, the density and viscosity of all DESs in this study decreased while their conductivity increased. Saputra et al. [39] investigated the thermophysical properties of highly stable novel ammonium-based ternary deep eutectic solvents (TDESs) by combining glycerol and two kinds of HBAs (ethyl ammonium chloride and zinc chloride), and a similar trend to the binary DESs in the previous work [25] was observed. It was because of the adjustable thermal stability of DESs, either binary or ternary, that made them suitable for processing composite resins in some specific conditions.

The investigation of the toxicity of DESs is indispensable for the assessment of safety, health and environmental impacts, and the toxicity of DESs must be taken into consideration before applying them in the processing of new materials. It was found that the phosphonium-based DESs were toxic on bacteria, while no toxicity was observed for ammonium-based DESs [40,41]. Hayyan et al. [42] investigated the cytotoxicity of ammonium-based DESs on five human cancer cell lines and one normal cell line. The DESs had inhabitation effect on cancer cell growth at certain condition and suggested that the toxicity of DESs may be related to the type of HBD and molar ratio of HBD/salts. Torregrosa-Crespo et al. [43] monitored the toxicity of DESs to *Escherichia coli*, and no toxic effect was observed at low DES concentration. When the concentration of DES was raised, the toxicity effect became obvious, which was considered due to the dual effect of both chemical composition of the DES and the high acidification of the media caused by DES hydrolysis during cellular growth. Wen et al. [44] assessed the toxicity of ChCl-based DESs comprising ChCl and choline acetate (ChAc) as the salts and urea, acetamide, glycerol and ethylene glycol as the HBDs on different living organisms. The effect of DESs on different living organisms was considered to be associated with their interactions with the cellular membranes. The toxic investigation on DESs provides meaningful ecological basis for the applications of DESs in the processing of composite resins.

3. Composite Resins Processing with DES

Since Type III DES formed through hydrogen bond interactions processed conveniently adjustable properties on hydrophobicity/hydrophilicity, viscosity, conductivity and thermal stabilities, the applications of DESs in the green processing of composite resins have attracted the interest of many researchers worldwide [21,45]. Table 1 showed the applications of DESs in various composite resin processing, which is nowadays a hot research area with a focus on the literature published in recent years.

Table 1. Summary of applications of deep eutectic solvents (DESs) in composite resin processing.

Deep Eutectic Solvent			Resin	Reference
Hydrogen-Bond Donor (HBD)	Hydrogen-Bond Acceptor (HBA)	Molar Ratio (HBD:HBA)		
Urea	Choline chloride	2:1	Epoxy resin (silane-functionalized epoxy resin)	[46]
Imidazole	Choline chloride	1:1	Epoxy resin (bisphenol A-based low molecular weight epoxy resin)	[47]
$ZnCl_2/SnCl_2$	Choline chloride	2:1	Epoxy resin (bisphenol A-based low molecular weight epoxy resin)	[48]
$SnCl_2$	Choline chloride	2:1	Epoxy resin (bisphenol A-based low molecular weight epoxy resin)	[48]
Aromatic amines (MPDA, DAT)	Choline chloride	2:1	Epoxy resin (bisphenol A-based low molecular weight epoxy resin)	[49]
Glycerol	Choline chloride	2:1	Epoxy resin (bisphenol F epoxy resin)	[50]
Ethylene glycol	Choline chloride	2:1	Epoxy resin (bisphenol F epoxy resin)	[50]

Table 1. Cont.

Deep Eutectic Solvent			Resin	Reference
Hydrogen-Bond Donor (HBD)	Hydrogen-Bond Acceptor (HBA)	Molar Ratio (HBD:HBA)		
Oxalic acid	Choline chloride	1:1	Epoxy resin (bisphenol F epoxy resin)	[50]
Tris(hydroxymethyl)propane	Choline chloride	1:1	Epoxy resin (bisphenol A-based low molecular weight epoxy resin)	[51]
Urea	Choline chloride	2:1	Epoxy resin (waterborne epoxy emulsion)	[52]
Urea	$ZnCl_2$	10:3	Phenolic resin	[53]
$ZnCl_2$	Acetamide	1:3	Phenolic resin	[54]
$ZnCl_2$	Choline chloride	2:1	Phenolic resin	[55]
Urea	$ZnCl_2$	1:1	Phenolic resin	[56]
Urea	$ZnCl_2$	10:3	Phenolic resin	[57]
Itaconic acid	Choline chloride	1:1	Acrylic resins	[58]
Methacrylic acid	Choline chloride	2:1	Acrylic resins	[59]
Acrylic acid	Choline chloride	1.6/2:1	Acrylic resins	[60]
Acrylic acid	Choline chloride	1.6/2:1	Acrylic resins	[60]
Acrylic acid	Lidocaine hydrochloride	3:1	Acrylic resins	[61]
Acrylic acid	Choline chloride	1.6/2:1	Acrylic resins	[62]
Acrylic acid	Choline chloride	1.6/2:1	Acrylic resins	[63]
Acrylic acid	Choline chloride	2:1	Acrylic resins	[64]
Acrylic acid	Benzalkonium chloride	2:1	Acrylic resins	[23]
Urea	Choline chloride	2:1	Polyester resin (polyethylene terephthalate)	[65]
Ethylene glycol	Choline chloride	2:1	Polyester resin (polyethylene terephthalate)	[66]
Glycerol	Choline chloride	2:1	Polyester resin (polyethylene terephthalate)	[67]
Ethylene glycol	Potassium carbonate	6:1	Polyester resin (polyethylene terephthalate)	[68]
1,3-dimethylurea	Zinc acetate	4:1	Polyester resin (polyethylene terephthalate)	[69]
Urea	Zinc chloride	4:1	Polyester resin (polyethylene terephthalate)	[70]
Choline chloride	Zinc acetate	1:1	Polyester resin (polyethylene terephthalate)	[71]
Urea	Choline chloride	2:1	Polyester resin (polyethylene terephthalate)	[72]
$ZnCl_2$	Choline chloride	2:1	Polyester resin (polyethylene terephthalate)	[72]
Methanesulfonic acid	Guanidine 1,5,7-triazabicyclo [4.4.0] dec-5-ene	1.5:0.1	Polyester resin (Polycaprolactone)	[73]
1,4-butanediol	3-(4-(4-(bis(2chloroethyl)amino) phenyl)butanoyloxy)-N,N,N-trimethylpropane-1-aminium chloride	6/5:1	Polyester resin (Polycaprolactone)	[74]
1,8-octanediol	Tetraethylammonium bromide	3:1	Polyester resin (octanediol-co-citrate polyesters)	[75]
1,8-octanediol	Hexadecyltrimethylammonium bromide	3:1	Polyester resin (octanediol-co-citrate polyesters)	[75]

Table 1. Cont.

Deep Eutectic Solvent			Resin	Reference
Hydrogen-Bond Donor (HBD)	Hydrogen-Bond Acceptor (HBA)	Molar Ratio (HBD:HBA)		
1,8-octanediol	Methyltriphenylphosphonium bromide	3:0.75	Polyester resin (octanediol-co-citrate polyesters)	[75]
Acetamide	Caprolactam	1:1	Polyester resin (methyl methacrylate)	[76]
Acetamide	Ammonium thiocyanate	3:1	Polyester resin (methyl methacrylate)	[76]
Ethylene	Tetrabutylammonium bromide	2:1	Polyester resin (methyl methacrylate)	[76]
Acrylic acid	Benzalkonium chloride	2:1	Polyester resin (ethoxylated bisphenol-a-glycidyl methacrylate and urethane dimethacrylate)	[77]
Caffeic acid	Choline chloride	0.4:1:1	Imprinted resin	[78]
Ethylene glycol	Choline chloride	0.4:1:1	Imprinted resin	[78]
Glycerol	Choline chloride	2:1	Imprinted resin	[79]
Ethylene glycol	Choline chloride	2:1	Imprinted resin	[80]
Ethylene glycol	Choline chloride	2:1	Imprinted resin	[81]
Ethylene glycol	Choline chloride	1:1	Imprinted resin	[14]
Glycerol	Choline chloride	1:1	Imprinted resin	[14]
Propylene glycol	Choline chloride	1:1	Imprinted resin	[14]
Glycerol	Allyl triethylammonium chloride	1:1	Imprinted resin	[82]

3.1. Epoxy Resin

Structural adhesives are widely used in the repair of degraded concrete or bonding fiber reinforced laminates to strengthen the concrete components. Among high-performance polymers used for the formulation of structural adhesives, epoxy resins are one of the most important materials to enhance their mechanical properties, adhesion performances and longevity [83,84]. However, the production of epoxy resins usually starts with bisphenol-A, a harmful compound to the human reproductive system, which also affects the curing kinetics when preparing epoxy resins [22,29,85]. Chen et al. [86] prepared a protocatechuic acid and epichlorohydrin-based epoxy resin instead of bisphenol-A-based, which had higher transition temperature than that of a cured commercial bisphenol-A epoxy resin, and its coefficient of thermal expansion was much lower. Hsu et al. [87] reported an epoxy resin reinforced with citric acid-modified cellulose (CAC) with improved tensile strength, Young's modulus and toughness than either pure epoxy resins or cellulose/epoxy composites. Herein, CAC was dispersed homogeneously in the composite, and the combination of terminal carboxyl groups of CAC and epoxy resin had a positive influence on the curing behavior of epoxy system in their work. Biomass had been proved to be an acceptable alternative to bisphenol-A, but it had limited functionalities to facilitate the curing reactions when preparing epoxy resins.

ILs have played multiple roles in the development of polymers, and the number of IL-based polymers has been increasing steadily [88]. ILs have been utilized in polymer science as curing agents to contribute to polymerization and additives to modify certain properties of polymers. Carvalho et al. [89] developed a new IL (1-butyl-3-methyl imidazolium tetrafluoroborate)-based epoxy resin used for coating. Compared with epoxy resins cured with conventional hardeners (anhydride or aromatic amine), the thermal stability of epoxy resins cured with the IL was found to be better. It was suggested that side reaction and plasticizing effect of IL may decrease the Tg of epoxy resins, which is unfavorable in the cases that high thermal stability of epoxy resins are required. Nguyen et al. [90] reported a new way to synthesize epoxy resins using a variety of phosphonium-based ILs and ILs displayed a high reactivity, curing properties to epoxy resins in all cases. The excellent thermal stabilities obtained in their study was attributed to the application of ILs as plasticizer in the curing of epoxy resins. Sidorov et al. [91] reported the chemical structures of some imidazolium and pyrrolidinium-based ILs and their influences on the curing behavior of epoxy resins, and it was shown that ILs possessed high catalytic activities without affecting the physical and chemical properties of current epoxy resins. Donato et al. [92] studied several epoxy composite resins using ILs (1-decyl-3-methylimidazolium tetrafluoroborate, 1-triethylene glycol monomethyl ether-3-methylimidazolium tetrafluoroborate and 1-triethylene glycol monomethyl ether-3-methylimidazolium methanesulfonate) as additives to control the interphase interaction in order to produce a fine dispersed morphology. It was indicated in their work that the thermomechanical properties of epoxy resins were improved due to the strong hydrogen bonds within the epoxy network.

The application of ILs in the curing of epoxy resins has paved a way to DES for processing of epoxy resins with green chemistry. Generally, metal chlorides are known as cationic catalysts for epoxy resins, most which are difficult to handle due to their cytotoxicity and instability in the water. Doolan et al. [93] found that $ZnCl_2$ and $SnCl_2$ exhibited lower cytotoxicity compared to other metal halides and preparing them into DESs may take use of their environmentally benign advantages [40]. Maka et al. [48] prepared $ZnCl_2$ and $SnCl_2$ into the form of DES by mixing them with a certain molar ratio of ChCl and used DESs (ChCl/$ZnCl_2$ and ChCl/$SnCl_2$) as curing cationic catalysts in the epoxy resins. It was observed that using DES of ChCl/$SnCl_2$ could produce higher curing efficiency for the epoxy resin than using DES of ChCl/$ZnCl_2$. Aromatic amines are another important class of curing agent used in the processing of epoxy resins. When mixing aromatic amines with epoxy resin, an elevated temperature is typically required to lower the viscosity of the mixture in order to get better homogeneity. M-phenylenediamine (MPDA) is one of the most commonly used aromatic amines in the curing of epoxy resin and Staciwa et al. [49] mixed it with ChCl to get a DES. After making MPDA into DES, the mixing process was made convenient because the viscosity of DES was lower than that of epoxy resin at the same temperature. Maka et al. [94] prepared guanidine-derived DESs (ChCl/guanidine thiocyanate (GTC) and ChCl/1-(o-tolyl) biguanide (TBG)). By using GTC and TBG as curing agent for the polymerization of the epoxy resin, the pot life values of the final products were shortened compared to that cured with ChCl/urea DES, and the viscosity jump because of the eutectic phenomenon was considered as the main affecting factor. The eutectic phenomenon is providing a hint for the processing of aromatic amines or epoxy resin curing agents with similar chemical structures, at lower temperature. In addition to the facilitation in lowering viscosities, preparing a certain component into DES can also decrease the interparticle aggregation and get better morphology control, which provided a new approach for the preparation of non-aqueous sol–gel-based epoxy resins. Lionetto et al. [95] studied a ChCl/urea-based epoxy resin and found that the addition of DESs can lower the Tg of epoxy resins and contribute to the dispersion of materials. Although in recent years, many researchers have found that DESs can replace the ILs in some applications, the use of DESs still has certain limitations. Mka et al. [47] investigated epoxy resins crosslinked with IL of 1-ethyl-3-methylimidazolium chloride ([EMIM] Cl) and DES of imidazole/ChCl, and they found that epoxy resins cured with IL exhibited more elastic properties and better mechanical features than those cured with DES.

3.2. Phenolic Resin

Phenolic resin is a type of product obtained via the condensation polymerization of acid compounds and aldehyde compounds. Optimizing the reactivity and heat resistance of phenolic resins is urgently needed in order to meet the requirements of production and manufacturing. From Figure 2, lignin has functional groups such as phenolic hydroxyl groups and alcoholic hydroxyl groups, which was similar to the chemical structure of phenols. Therefore, a similar method for modifying lignin can be used for the modification of phenolic resins. Taking the advantages of DESs into consideration, the lignin modified by DESs could also be used for the modification of phenolic resins [96,97]. Cao et al. [55] reported that lignin could replace phenol to prepare phenolic resins, and lignin in their work was modified with DES (ChCl/ZnCl$_2$ = 1:2) to improve the reactivity. By analyzing the structural change of lignin before and after modification, modified lignin may undergo partial demethylation reaction compared with lignin. The methoxy group on its aromatic ring was oxidized to phenolic hydroxyl group under the condition of Lewis acid, forming a small amount of active catechol structure. Although the bond strength of the DES-modified lignin used to prepare phenolic resins decreased due to the limited amount of modified lignin used for phenolic resins, the bond strength of phenolic resins could still reach the national standard. Lian et al. [53] introduced ZnCl$_2$ and urea based on DES into phenolic resin and an increase in Tg of phenolic resin was reported with the increase of Zn content. The onset of thermal degradation of modified phenolic resins (filled with relignin) occurred at higher temperatures than those resins with unmodified lignin and without it, showing superior thermal properties. DESs impacted the interaction between the intermolecular and intramolecular in the lignin network due to the action of the ionic environment, and the chemical structure of lignin was found to be influenced by Zn by comparing the FTIR spectra of lignin and relignin, where a similar conclusion was drawn by Sun et al. [98]. Hong et al. [54] compared the phenolic resins modified by lignin (LPF), DES/lignin (DLPF) and DES-modified lignin (MLPF) and reported that the curing time was in a trend of DLPF > MLPF > LPF. MLPF exhibited higher bond strength and shorter curing time than the phenolic resin. It indicated that modification with DES was the most efficient approach to improve the mechanical, reactivity and thermal properties of phenolic resin in their study.

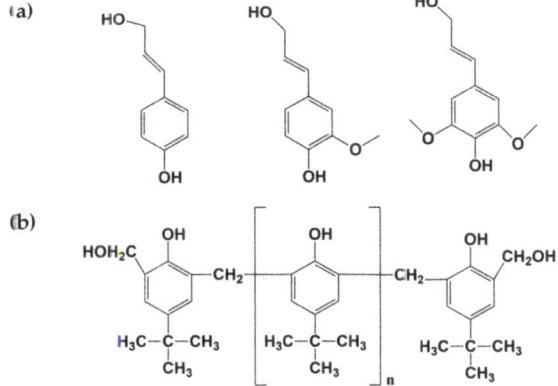

Figure 2. Chemical structure of (a) monolignol and (b) phenolic resin.

Nowadays, carbonaceous materials have been widely used for various purposes, such as catalysis, gas separating or capture and electrode of supercapacitor, and phenolic resin-based carbon is one of the most studied carbonaceous materials. Due to pollution and recovery issues, surfactants or block-copolymers are no longer suitable as structure directing agents for carbon materials with controlled pore characteristics [99,100]. Deng et al. [57] have used DESs instead of conventional pore structure directing agents to reduce pollution and solve recovery problems. The phenolic resin was selected as the raw material, which was mixed with DES composed of ZnCl$_2$ and urea to obtain a carbonaceous material. Compared to the traditional ways, DES-based carbonaceous materials could drastically decrease the time of the whole manufacturing process, increase the specific surface area and optimized electrochemical properties. Moreover, Zhong et al. [56] used DES (urea/ZnCl$_2$ = 1:1) and phenolic resins as raw materials via direct carbonization process to manufacture nanoporous carbon materials. The results showed that modified nanoporous carbon materials had large specific capacitance, superior cycling stability and excellent energy density in alkaline KOH electrolyte. It is a commendable idea to use DESs as new green solvents to assist in the synthesis of carbonaceous materials or to modify carbonaceous materials.

3.3. Acrylic Resins

Acrylic resins are a well-known class of pH-responsive polymers that have been used as absorbers, anti-fouling agents and carriers, and they have excellent resistance to weather, heat and chemical. To date, researchers from worldwide are dedicated on the issues corresponding to raw materials, production technology and product quality of acrylic resins [101–103]. DESs exhibit similar properties to ILs that make them suitable for the processing of acrylic resins, which not only prevents the use of VOCs, but also offers a green approach for the functionalization of acrylic resins.

The design of acrylic acid (AA) and methacrylic acid (MA)-based DESs paved the way for the development of frontal polymerization in which DESs were first used as monomers. Fazende et al. [59] prepared a series of acrylic resins with DESs of acrylic acid (AA) and methacrylic acid (MA) as HBDs as well as ChCl as HBA through frontal polymerization and the talc, dimethyl sulfoxide, lauric acid and stearic acid were used to replace ChCl to determine the impact of DESs in acrylic resins polymerization. It was observed that the polymerization in the DESs was well controlled and evenly propagated, and the introduction of ChCl increased the chemical reactivity of AA and MA. Mota-Morales et al. [60] further took advantage of the capability of DESs as solvents to disperse carbon nanotube and reported the preparation of poly acrylic acid (PAA)-carbon nanotube composites by incorporating carbon nanotube in the polymerizable DES (AA/ChCl), and carbon nanotube played a role of filler in the acrylic resin. A desired macroporosity of the acrylic resin was formed due to the strong interaction between carbon nanotube and PAA. Furthermore, Mota-Morales et al. [61] reported a series of acrylic resins based on the DESs which were formed through hydrogen bonds of the acrylic monomers and ammonium salts. These DESs played an all-in-one role, such as monomer, solvent and fillers, in acrylic resins, and the adjustment of viscosities and double bonds in DESs favored the extraordinary conversion of the monomers by frontal polymerization. In their work, they prepared DES acrylic resin loaded with lidocaine hydrochloride (ammonium salts) as HBA and it was observed that the conversion rate of AA was 100% at a mild temperature and the release of lidocaine hydrochloride could be controlled conveniently utilizing the swelling behavior of PAA affected by pH. A similar explanation on kinetics was reported by Sánchez-Leija et al. [104], and it was claimed that the release of lidocaine hydrochloride was controlled by multiple factors of pH, ionic strength and solubility in the polymer coupled with the swelling of polymers and the specific interactions between the lidocaine hydrochloride and the polymer.

DESs were also applied for acrylic resin-based functional polymers to give novel properties to meet certain requirements. Li et al. [62] reported a flexible tactile/strain sensor based on the photopolymerization of AA/ChCl DES with favorable transparency (transmittance of ~81%), elasticity (strain up to 150%) and conductivity (~0.2 S m^{-1}). The transparency of the materials was able to be triggered by complex cross-linked molecular networks. The elasticity was ascribed to the flexible macromolecular chains consisted of easily changing molecular configuration of AA parts within the DES, while the conductivity was owing to the moving positive/negative ions of ChCl parts within the DES. Moreover, the sensor was sensitive to an external strain which may be due to weak hydrogen bond contact between ChCl and AA. Li et al. [63] further studied a green fabrication of conductive paper with stable electrical conductivity and increased elasticity by polymerization of ChCl and AA-based DES. An increase on the elasticity of conductive paper was observed, and it was considered due to the poly DES (acrylic resin) elastomers bridging neighboring cellulose fibers. Similarly, Wang et al. [64] reported a novel transparent and conductive wood using renewable wood as substrate and polymerizable DES of ChCl and AA as backfilling agent. The cellulose orientation and strong interactions between the cellulose template and the polymerizable DES contributed to excellent stretch ability of the wood. Wang et al. [23] reported two acrylic resins synthesized with polymerizable DESs, which were designed by using benzalkonium chloride (BC) as the HBA and AA/MA as the HBDs. The obtained acrylic resins exhibited not only adjustable flexibility, but also antimicrobial properties. In this work, BC within the DES, as an antimicrobial agent, was in charge of antimicrobial functionality, while AA or MA, two commonly used unsaturated acrylic acid, performed the polymerization reaction. The properties and functions of DESs can be easily tailored through the selection of HBDs and HBAs and tuning on their molar ratios, which provided more possibilities for the processing of acrylic resins [105]. Moreover, the uses of DESs as monomers and polymerization medium in acrylic resins have become a spotlight, and it is believed that the multiples roles of DESs in the processing of acrylic resins should be highlighted.

3.4. Polyester Resin

Polyester resin is a kind of polymer containing unsaturated bonds. It is conventionally processed by two routes: (1) transesterification and polycondensation of dimethyl ester and diol to form polymer; (2) esterification of the diacid and diol and then polycondensation [106]. It is a relatively important material used in manufacturing fibers, concrete and coating [107,108]. Researchers have modified polyester resins by various means, taking the factors of mechanical performance, environmental protection and biocompatibility into consideration.

Poly (ethylene terephthalate) (PET) is one of the most important polyesters widely used in the manufacturing of plastic bottles and fabrics. PET bottles recycling and waste management have become an especially critical issue for environmental protection. Modifications of PET are helpful to improve the fiber suitability to be used as fabrics. There is a trial stage for DESs to take charge in the modifications taking advantage of their design flexibility. ChCl/urea-based DES [65] and ChCl/ethylene glycol-based DES [66] have been tried for hydrophilicity/hydrophobicity surface modifications of PET with microwave irradiation. Researchers have found that DES was an ideal solvent that can not only functionalize the PET surface with desired properties, but also help avoid side-reactions with more environmentally benign results. Depolymerization of PET is one of the easiest approaches to deal with PET; DESs have been utilized to facilitate the processing. Glycolysis of PET is an effective approach to transform PET back to its monomer of bis(hydroxyethyl)terephthalate (BHET) after recycling in order to remake qualified products, and DES has played multiple roles to make the process greener. Liu et al. [109] explored the possibility and processing efficiency of glycolysis of PET with IL of choline acetate ([Ch][OAc]), and they proposed that the enhancement of the glycolysis reaction was due to the formation of hydrogen bonds between PET and the IL. The formation of hydrogen bonds is a notable characteristic of Type III DES, so replacing ILs with DES for the glycolysis of PET is considered feasible in theory. Choi et al. [67] have utilized DES of ChCl/glycerol to replace conventional solvent used in PET depolymerization, with the assistance of microwave irradiation, the processing of PET was simplified and improved in energy efficiency. Sert et al. [68] synthesized five DESs to catalyze the glycolysis of PET, and the DES of potassium carbonate and ethylene glycol was considered to be the most efficient catalyst with highest monomer product yield. Liu et al. [69] also explored the possibility to use DES of 1,3-dimethylurea/Zn (OAc)$_2$ for PET glycolysis, and they attributed the high catalyze efficiency to the synergistic effect of acid and based formed within DES. A similar result was obtained by Wang et al. [70] when using DES of urea/ZnCl$_2$ to catalyze the glycolysis of PET, in which the role of hydrogen bonds within DES was addressed. Instead of obtaining BHET, Zhou et al. [71] utilized a DES of ChCl/Zn (Ac)$_2$ in the alcoholysis of PET to produce dioctyl terephthalate (DOTP), which is a plasticizer used in polymer industry. They also found that the hydrogen bonds within DES were important to accelerate the degradation process. Aminolytic depolymerization of PET has provided other ways to recycle and use of the waste plastic bottles. Musale et al. [72] proved that DESs of ChCl/urea and ChCl/ZnCl$_2$ are efficient catalysts for the aminolytic depolymerization of PET, yielding pure products of N_1, N_1, N_4, N_4-tetrakis (2-hydroxyethyl)-terephthalamide (THETA) of 82%, terephthalic acid (TPA) of 83% and bis (2-hydroxy ethylene) terephthalamide (BHETA) of 95%, respectively.

Some other polyesters may include polybutylene terephthalate (PBT), polyethylene naphthalate (PEN), polycaprolactone (PCL) and polyacrylate (PAR). Rare published works were found concerning the processing of PBT and PEN with DES, but more frontier research has been conducted regarding the processing of PCL and PAR. García-Argüelles et al. [73] reported that polycaprolactones (PCLs) was able to be synthesized by using DES of methane sulfonic acid (MeSO$_3$H) and the guanidine 1,5,7-triazabicyclo [4.4.0] dec-5-ene (TBD) as catalyst, where DES replaced other solvents and initiators minimizing the presence of harmful chemical reagents. PCL is sometimes used as carrier for drug deliveries and Pradeepkumar et al. [74] designed a DES-mediated drug carrier where DES influenced the formation of folic acid (FA)-tagged g-β-alanine-co-PCL polymer (DES@FA-g-β-alanine-co-PCL). This amphiphilic polymer is expected to have great potential for pharmaceutical applications owing to the controlled release of drug from DES@FA-g-β-alanine-co-PCL as the carrier. Functionalization of polyesters is attractive to researchers for expanding their applications; antimicrobial properties are favorable when polyesters were used for biomedical purposes. Zhou et al. [110] added quaternary ammonium compound (QAC) as antibacterial agent to synthesize a new kind of polyester and an excellent antibacterial effect for *Escherichia coli* and *Staphylococcus aureus* were observed. However, directly adding QACs to polyesters may have the problem of burst release due to the lack of bonding between QACs and the polymer network, which greatly affect the long-term antimicrobial efficiency of the modified polyester [111]. García-Argüelles et al. [75] prepared the antimicrobial agents of quaternary ammonium or phosphonium salt into the form of DES by using 1,8-octanediol as HBA and acquired antibacterial properties were obtained with stability and biocompatibility. Hydrogen bonds between the halide anion of quaternary nitrogen or phosphonium salt and the hydroxyl groups in 1,8-octanediol were considered as the main reason to stabilize the antimicrobial groups within the polyester network. Wang et al. [76] studied the polymerization of methyl methacrylate by metal Fe catalysts with three types of DESs as solvents, or ligands. The polymerization process could be well controlled and more environmentally friendly. DES could also be easily separated from the polymerization system. Wang et al. [77] incorporated DES (benzalkonium chloride/acrylic acid) into a dental composite resins with PARs as the main components, and the incorporation of DES produced better maintenance of the mechanical properties, biocompatibility than adding benzalkonium chloride as antimicrobial agent directly. Hydrogen bond within DES was considered as the main factor that contribute to high mechanical strength and limited releasing of BC within the polymer network. Proper utilization of the eutectic phenomena and the hydrogen bonds within DES is inspiring the processing of polyester resins for a variety of future applications.

3.5. Imprinted Resin

Molecularly imprinted polymers (MIPs), also known as "plastic antibodies", is a new research area developed based on molecular recognition theory that has been growing rapidly in recent years. A typical approach to prepare MIP may include the following steps [82,112–114]: (1) synthesizing a complex through non-covalent bonds or covalent methods under a certain temperature or pressure with a template and the functional monomer; (2) cross-linking the functional monomer through polymerization reaction with the formed polymer surrounding the template; (3) removing the template from the polymer by a specific method with MIP remained. For the moment, MIPs can selectively recognize the template molecules or their analogs though the specific recognition site or the three-dimensional cavity with a spatially complementary structure formed with the template molecule and the functional monomer during preparation (Figure 3). MIPs have shown broad application prospects in molecular recognition, purification and material processing and have become a research hotspot in the field of polymeric materials [115,116]. In order to solve environmental pollution, high costs and hydrophilic issues of traditionally imprinted polymers, DESs have been applied to the molecular imprinting processing.

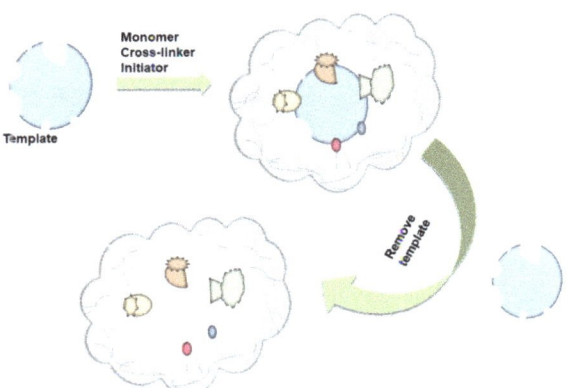

Figure 3. Procedure for preparing molecularly imprinted polymers (MIPs).

The monomers and templates in MIPs are generally two separate constitutions, which may lead to complications in processing and high costs for MIPs polymerization. The combination of both can greatly improve production efficiency and save resources. Fu et al. [78] reported a ternary DES of ChCl/caffeic acid/ethylene glycol that played dual roles of template and functional monomer in the preparation of MIPs. It was illustrated in their work that MIPs prepared with the ternary DES had a good recognition ability for polyphenols, and this MIP was stable at room temperature because of the increased recognition sites between template and monomer polymer. The choice of the elution solution in MIPs is also of great importance, because it may significantly affect the production efficiency and manufacturing costs. Li et al. [79] prepared MIPs with caffeic acid as a template and a mixture of methanol and DES (ChCl/glycerol = 1:2) as elution solution, which were used for rapid purification of caffeic acid from hawthorn. The template could be extracted easily by the elusion solution according to the "like dissolves like" theory, which provided a new approach to improve the preparation efficiency through the choice of elution solution. Other drawbacks of MIPs may include poor compatibility and molecule-recognition ability in the aqueous phase—especially in the cases where various aqueous biologic and environmental samples require MIPs to be compatible with aqueous media. The introduction of hydrophilic groups into MIPs is an effective approach to adjust the hydrophilicity/hydrophobicity of MIPs. Tang et al. [80] introduced hydrophilic resorcinol and melamine as monomers and DESs (ChCl/ethylene glycol, tetramethylammonium bromide/ethylene glycol and tetramethylammonium chloride/ethylene glycol) as solvents into MIPs targeting to recognize quinolones (ciprofloxacin and ofloxacin), and the hydrophilicity of MIPs was increased owing to the hydrophilic groups from both the monomers and DESs. It was observed that the MIPs prepared with DESs—especially the DES of ChCl and ethylene glycol—had excellent recoveries and purification of quinolones in wastewater. Liang et al. [81] synthesized molecularly imprinted phloroglucinol–formaldehyde–melamine resin (MIPFMR) using a hydrophilic resin and MIPs technology in the DES (ChCl/ethylene glycol = 1:2) to enhance the affinity of the MIPFMR for analytes in aqueous media. MIPFMR was prepared by using phenylephrine phloroglucinol as dummy template, melamine as bifunctional monomer and formaldehyde as cross-linker. Hydroxyl groups, ether linkages and amino groups were able to be introduced to MPIFMR by using phloroglucinol and melamine, which could interact with the template in DES via hydrogen bonds and π–π interactions. Moreover, MIPFMR prepared in DES showed higher adsorption property and hydrophilicity than resins prepared in alcoholic solvent systems, because polar solvents were not favorable to the formation of hydrogen bonds between templates and monomers. There have been many studies to introduce DESs in the preparation of MIPs for modification, purification and functionalization, but the expansions of the roles of DESs in the processing of MIPs still need further exploration by researchers.

4. Conclusions and Outlook

Composite resins are widely used in public transportation, construction supplies, biomedicine and industrial productions. Advancement in processing of composite resins to make them better suited for certain applications is of great interest to both academic and industrial communities. DESs have been found to be a greener and more economical alternative to conventional organic solvents or ILs and can be easily prepared by mixing two or more components through the formation of hydrogen bonds. This review addressed the development of DESs in the processing of composite resins. For example, DESs can be used for the synthesis and modifications of various polymeric materials. Different DESs can be also applied to a certain type of composite resin to meet specific requirements. Previous successful practices have demonstrated the feasibility of using DESs as solvents, monomers and catalysts in polymerization reactions, which has provided new clues for the development of resin preparations. Taking advantage of the designable properties of DESs through the selection of HBDs and HBAs, it has been possible to adjust the physiochemical properties of DESs to better match certain applications during the processing of resin composite. Previous works have shown that DESs can play multiple roles in the processing of functional composite resins, and DESs still hold great potentials in the development of polymer sciences.

The development of DES-based resin composites can be a promising approach for further improvement of resin composites. In addition, profound knowledge of resins and DESs are keys to better understand the behavior of resin composites in certain applications. However, due to the limitations of the design principles of DESs, it has been difficult to gain a clear understanding on the intermolecular forces when DESs are introduced to a new polymeric matrix. Therefore, in-depth research is necessary to clarify the interactions between DESs and the original resin composites—especially in terms of the functionality of hydrogen bonds within DESs in the complexed composite resins. This review aimed to provide a summary of previous works towards the processing of composite resins with DESs, and hopefully an inspiration of low-cost technology to high-tech products was able to be made.

Author Contributions: Writing—original draft preparation, J.W., J.X., D.F.; writing—review and editing, H.H., M.W.; funding acquisition, J.W. All authors have read and agreed to the published version of the manuscript.

Funding: This work was funded by the University City Fusion Project from Science and Technology Bureau of Zhangdian District (Zibo, China), Grant Number 9001-118246 (J.W.) and the startup fund from Shandong University of Technology (J.W. and M.W.).

Acknowledgments: J.W. and M.W. would like to specially acknowledge the startup fund from Shandong University of Technology that were used in partial support of this work.

Conflicts of Interest: The authors declare no conflict of interest.

References

1. Post, W.; Susa, A.; Blaauw, R.; Molenveld, K.; Knoop, R.J.I. A Review on the potential and limitations of recyclable thermosets for structural applications. *Polym. Rev.* **2020**, *60*, 359–388. [CrossRef]
2. Cabanes, A.; Valdés, F.J.; Fullana, A. A review on VOCs from recycled plastics. *Sustain. Mater. Technol.* **2020**, *25*, e00179. [CrossRef]
3. Bhattacharya, A. Grafting: A versatile means to modify polymers techniques, factors and applications. *Prog. Polym. Sci.* **2004**, *29*, 767–814. [CrossRef]
4. Abbott, A.P.; Capper, G.; Davies, D.L.; Rasheed, R.K.; Tambyrajah, V. Novel solvent properties of choline chloride/urea mixtures. *Chem. Commun.* **2003**, *9*, 70–71. [CrossRef] [PubMed]
5. Abbott, A.P.; Boothby, D.; Capper, G. Deep eutectic solvents formed between choline chloride and carboxylic acids: Versatile alternatives to ionic liquids. *J. Am. Chem. Soc.* **2004**, *126*, 9142–9147. [CrossRef] [PubMed]
6. Zhang, Q.; De Oliveira Vigier, K.; Royer, S.; Jerome, F. Deep eutectic solvents: Syntheses, properties and applications. *Chem. Soc. Rev.* **2012**, *41*, 7108–7146. [CrossRef] [PubMed]
7. Tang, W.; Row, K.H. Hydrophobic ionic liquid modified thermoresponsive molecularly imprinted monolith for the selective recognition and separation of tanshinones. *J. Sep. Sci.* **2018**, *41*, 3372–3381. [CrossRef]

8. Tang, W.; Ho Row, K. Evaluation of CO_2-induced azole-based switchable ionic liquid with hydrophobic/hydrophilic reversible transition as single solvent system for coupling lipid extraction and separation from wet microalgae. *Bioresour. Technol.* **2020**, *296*, 122309. [CrossRef]
9. Abbott, A.P.; Capper, G.; Davies, D.L. Solubility of metal oxides in deep eutectic solvents based on choline chloride. *J. Chem. Eng. Data* **2006**, *51*, 1280–1282. [CrossRef]
10. Khandelwal, S.; Tailor, Y.K.; Kumar, M. Deep eutectic solvents (DESs) as eco-friendly and sustainable solvent/catalyst systems in organic transformations. *J. Mol. Liq.* **2016**, *215*, 345–386. [CrossRef]
11. Shishov, A.; Pochivalov, A.; Nugbienyo, L.; Andruch, V.; Bulatov, A. Deep eutectic solvents are not only effective extractants. *Trac Trend Anal. Chem.* **2020**, *129*, 115956. [CrossRef]
12. Vilková, M.; Płotka-Wasylka, J.; Andruch, V. The role of water in deep eutectic solvent-base extraction. *J. Mol. Liq.* **2020**, *304*, 112747. [CrossRef]
13. Lou, R.; Ma, R.; Lin, K.; Ahamed, A.; Zhang, X. Facile extraction of wheat straw by deep eutectic solvent (DES) to produce lignin nanoparticles. *ACS Sustain. Chem. Eng.* **2019**, *7*, 10248–10256. [CrossRef]
14. Tang, W.; Gao, F.; Duan, Y.; Zhu, T.; Ho Row, K. Exploration of deep eutectic solvent-based molecularly imprinted polymers as solid-phase extraction sorbents for screening chloramphenicol in milk. *J. Chromatogr. Sci.* **2017**, *55*, 654–661. [CrossRef]
15. Carriazo, D.A.; Gutierrez, M.C.; Ferrer, M.L.; del Monte, F. Deep-eutectic solvents playing multiple roles in the synthesis of polymers and related materials. *Chem. Soc. Rev.* **2012**, *41*, 4996–5014. [CrossRef]
16. Tang, B.; Row, K.H. Recent developments in deep eutectic solvents in chemical sciences. *Mon. Für Chem. Mon.* **2013**, *144*, 1427–1454. [CrossRef]
17. Francisco, M.D.; van den Bruinhorst, A.; Kroon, M.C. Low-transition-temperature mixtures (LTTMs): A new generation of designer solvents. *Angew. Chem. Int. Ed.* **2013**, *52*, 3074–3085. [CrossRef]
18. Ge, X.; Gu, C.; Wang, X.; Tu, J. Deep eutectic solvents (DESs)-derived advanced functional materials for energy and environmental applications: Challenges, opportunities, and future vision. *J. Mater. Chem. A* **2017**, *5*, 8209–8229. [CrossRef]
19. Smith, E.L.; Abbott, A.P.; Ryder, K.S. Deep eutectic solvents (DESs) and their applications. *Chem. Rev.* **2014**, *114*, 11060–11082. [CrossRef]
20. Abbott, A.P.; Frisch, G.; Garrett, H.; Hartley, J. Ionic liquids form ideal solutions. *Chem. Commun.* **2011**, *47*, 11876. [CrossRef]
21. Perna, F.M.; Vitale, P.; Capriati, V. Deep eutectic solvents and their applications as green solvents. *Curr. Opin. Green Sustain. Chem.* **2020**, *21*, 27–33. [CrossRef]
22. Sun, H.; Li, Y.; Wu, X.; Li, G. Theoretical study on the structures and properties of mixtures of urea and choline chloride. *J. Mol. Modeling* **2013**, *19*, 2433–2441. [CrossRef] [PubMed]
23. Wang, J.; Xue, J.; Dong, X.; Yu, Q.; Baker, S.; Wang, M.; Huang, F. Antimicrobial properties of benzalkonium chloride derived polymerizable deep eutectic solvent. *Int. J. Pharm.* **2020**, *575*, 119005. [CrossRef] [PubMed]
24. Gautam, R.; Kumar, N.; Lynam, J.G. Theoretical and experimental study of choline chloride-carboxylic acid deep eutectic solvents and their hydrogen bonds. *J. Mol. Struct.* **2020**, *1222*, 128849. [CrossRef]
25. Troter, D.; Đokić-Stojanović, D.; Đorđević, B.; Todorovic, V.; Konstantinović, S.; Veljković, V. The physicochemical and thermodynamic properties of the choline chloride-based deep eutectic solvents. *J. Serb. Chem. Soc.* **2017**, *82*, 1039–1052. [CrossRef]
26. Aroso, I.M.; Craveiro, R.; Dionisio, M.; Barreiros, S.; Reis, R.L.; Paiva, A.; Duarte, A.R.C. Fundamental studies on natural deep eutectic solvents: Physico-chemical, thermal and rheological properties. In Proceedings of the 6th Nordic Wood Biorefinery Conference, Helsinki, Finland, 20–22 October 2015; pp. 155–160.
27. Tomai, P.; Lippiello, A.; D'Angelo, P.; Persson, I.; Martinelli, A.; Di Lisio, V.; Curini, R.; Fanali, C.; Gentili, A. A low transition temperature mixture for the dispersive liquid-liquid microextraction of pesticides from surface waters. *J. Chromatogr.* **2019**, *1605*, 360329. [CrossRef]
28. Francisco, M.; van den Bruinhorst, A.; Kroon, M.C. New natural and renewable low transition temperature mixtures (LTTMs): Screening as solvents for lignocellulosic biomass processing. *Green Chem.* **2012**, *14*, 2153–2157. [CrossRef]
29. Choi, Y.H.; van Spronsen, J.; Dai, Y.; Verberne, M.; Hollmann, F.; Arends, I.W.C.E.; Witkamp, G.-J.; Verpoorte, R. Are natural deep eutectic solvents the missing link in understanding cellular metabolism and physiology? *Plant Physiol.* **2011**, *156*, 1701–1705. [CrossRef]

30. Stefanovic, R.; Webber, G.B.; Page, A.J. Polymer solvation in choline chloride deep eutectic solvents modulated by the hydrogen bond donor. *J. Mol. Liq.* **2019**, *279*, 584–593. [CrossRef]
31. Tiecco, M.; Cappellini, F.; Nicoletti, F.; Del Giacco, T.; Germani, R.; Di Profio, P. Role of the hydrogen bond donor component for a proper development of novel hydrophobic deep eutectic solvents. *J. Mol. Liq.* **2019**, *281*, 423–430. [CrossRef]
32. Tang, W.; Dai, Y.; Row, K.H. Evaluation of fatty acid/alcohol-based hydrophobic deep eutectic solvents as media for extracting antibiotics from environmental water. *Anal. Bioanal. Chem.* **2018**, *410*, 7325–7336. [CrossRef] [PubMed]
33. Tang, W.; Row, K.H. Design and evaluation of polarity controlled and recyclable deep eutectic solvent based biphasic system for the polarity driven extraction and separation of compounds. *J. Clean. Prod.* **2020**, *268*, 122306. [CrossRef]
34. Tang, W.; An, Y.; Row, K.H. Recoverable deep eutectic solvent-based aniline organic pollutant separation technology using choline salt as adsorbent. *J. Mol. Liq.* **2020**, *306*, 112910. [CrossRef]
35. Sas, O.G.; Fidalgo, R.; Domínguez, I.; Macedo, E.A.; González, B. Physical properties of the pure deep eutectic solvent, [ChCl]: [Lev] (1:2) DES, and its binary mixtures with alcohols. *J. Chem. Eng. Data* **2016**, *61*, 4191–4202. [CrossRef]
36. Abbott, A.P.; Capper, G.; Gray, S. Design of improved deep eutectic solvents using hole theory. *Chemphyschem* **2006**, *7*, 803–806. [CrossRef]
37. Chen, W.; Xue, Z.; Wang, J.; Jiang, J.; Zhao, X.; Mu, T. Investigation on the thermal stability of deep eutectic solvents. *Acta Phys. Chim. Sin.* **2018**, *34*, 904–911. [CrossRef]
38. Ghaedi, H.; Ayoub, M.; Sufian, S.; Lal, B.; Uemura, Y. Thermal stability and FT-IR analysis of Phosphonium-based deep eutectic solvents with different hydrogen bond donors. *J. Mol. Liq.* **2017**, *242*, 395–403. [CrossRef]
39. Saputra, R.; Walvekar, R.; Khalid, M.; Mubarak, N.M. Synthesis and thermophysical properties of ethylammonium chloride-glycerol-ZnCl2 ternary deep eutectic solvent. *J. Mol. Liq.* **2020**, *310*, 113232. [CrossRef]
40. Hayyan, M.; Hashim, M.A.; Hayyan, A.; Al-Saadi, M.A.; AlNashef, I.M.; Mirghani, M.E.S.; Saheed, O.K. Are deep eutectic solvents benign or toxic? *Chemosphere* **2013**, *90*, 2193–2195. [CrossRef]
41. Hayyan, M.; Hashim, M.A.; Al-Saadi, M.A.; Hayyan, A.; AlNashef, I.M.; Mirghani, M.E.S. Assessment of cytotoxicity and toxicity for phosphonium-based deep eutectic solvents. *Chemosphere* **2013**, *93*, 455–459. [CrossRef]
42. Hayyan, M.; Looi, C.Y.; Hayyan, A.; Wong, W.F.; Hashim, M.A. In vitro and in vivo toxicity profiling of ammonium-based deep eutectic solvents. *PLoS ONE* **2015**, *10*, 0117934. [CrossRef] [PubMed]
43. Torregrosa-Crespo, J.; Marset, X.; Guillena, G.; Ramón, D.J.; María Martínez-Espinosa, R. New guidelines for testing "Deep eutectic solvents" toxicity and their effects on the environment and living beings. *Sci. Total Environ.* **2020**, *704*, 135382. [CrossRef] [PubMed]
44. Wen, Q.; Chen, J.X.; Tang, Y.L.; Wang, J.; Yang, Z. Assessing the toxicity and biodegradability of deep eutectic solvents. *Chemosphere* **2015**, *132*, 63–69. [CrossRef]
45. Das, S.; Mondal, A.; Balasubramanian, S. Recent advances in modeling green solvents. *Curr. Opin. Green Sustain. Chem.* **2017**, *5*, 37–43. [CrossRef]
46. Lionetto, F.; Timo, A.; Frigione, M. Cold-cured epoxy-based organic-inorganic hybrid resins containing deep eutectic solvents. *Polymers* **2018**, *11*, 14. [CrossRef]
47. Maka, H.; Spychaj, T. Epoxy resin crosslinked with conventional and deep eutectic ionic liquids. *Polim. Polym.* **2012**, *57*, 34–40. [CrossRef]
48. Mąka, H.; Spychaj, T.; Adamus, J. Lewis acid type deep eutectic solvents as catalysts for epoxy resin crosslinking. *RSC Adv.* **2015**, *5*, 82813–82821. [CrossRef]
49. Staciwa, P.; Spychaj, T. New aromatic diamine-based deep eutectic solvents designed for epoxy resin curing. *Polimery* **2018**, *63*, 453–457. [CrossRef]
50. Gholami, H.; Arab, H.; Mokhtarifar, M.; Maghrebi, M.; Baniadam, M. The effect of choline-based ionic liquid on CNTs' arrangement in epoxy resin matrix. *Mater. Des.* **2016**, *91*, 180–185. [CrossRef]
51. Mąka, H.; Spychaj, T.; Kowalczyk, K. Imidazolium and deep eutectic ionic liquids as epoxy resin crosslinkers and graphite nanoplatelets dispersants. *J. Appl. Polym. Sci.* **2014**, *131*, 40401. [CrossRef]

52. Guo, L.; Gu, C.; Feng, J.; Guo, Y.; Jin, Y.; Tu, J. Hydrophobic epoxy resin coating with ionic liquid conversion pretreatment on magnesium alloy for promoting corrosion resistance. *J. Mater. Sci. Technol.* **2020**, *37*, 9–18. [CrossRef]
53. Lian, H.; Hong, S.; Carranza, A.; Mota-Morales, J.D.; Pojman, J.A. Processing of lignin in urea–zinc chloride deep-eutectic solvent and its use as a filler in a phenol-formaldehyde resin. *RSC Adv.* **2015**, *5*, 28778–28785. [CrossRef]
54. Hong, S.; Sun, X.; Lian, H.; Pojman, J.A.; Mota-Morales, J.D. Zinc chloride/acetamide deep eutectic solvent-mediated fractionation of lignin produces high- and low-molecular-weight fillers for phenol-formaldehyde resins. *J. Appl. Polym. Sci.* **2019**, *137*, 48385. [CrossRef]
55. Cao, X.; Hong, S.; Ding, Q.; Chen, F.; Zhu, M.; Lian, H. Lignin modified by DES with choline chloride/Zinc chloride to prepared phenol-formaldehyde resin adhesive. *China For. Prod. Ind.* **2016**, *43*, 25–29.
56. Zhong, M.; Liu, H.; Wang, M.; Zhu, Y.W.; Chen, X.Y.; Zhang, Z.J. Hierarchically N/O-enriched nanoporous carbon for supercapacitor application: Simply adjusting the composition of deep eutectic solvent as well as the ratio with phenol-formaldehyde resin. *J. Power Sources* **2019**, *438*, 226982. [CrossRef]
57. Deng, J.; Chen, L.; Hong, S.; Lian, H. UZnCl$_2$-DES assisted synthesis of phenolic resin-based carbon aerogels for capacitors. *J. Porous Mater.* **2020**, *27*, 789–800. [CrossRef]
58. Liu, Y.; Wang, Y.; Lai, Q.; Zhou, Y. Magnetic deep eutectic solvents molecularly imprinted polymers for the selective recognition and separation of protein. *Analytica Chimica Acta* **2016**, *93*, 168–178. [CrossRef]
59. Fazende, K.F.; Phachansitthi, M.; Mota-Morales, J.D.; Pojman, J.A. Frontal polymerization of deep eutectic solvents composed of acrylic and methacrylic acids. *J. Polym. Sci. Part A Polym. Chem.* **2017**, *55*, 4046–4050. [CrossRef]
60. Mota-Morales, J.D.; Gutiérrez, M.C.; Ferrer, M.L.; Jiménez, R.; Santiago, P.; Sanchez, I.C.; Terrones, M.; Del Monte, F.; Luna-Bárcenas, G. Synthesis of macroporous poly(acrylic acid)-carbon nanotube composites by frontal polymerization in deep-eutectic solvents. *J. Mater. Chem.* **2013**, *1*, 3970–3976. [CrossRef]
61. Mota-Morales, J.D.; Gutiérrez, M.C.; Ferrer, M.L.; Sanchez, I.C.; Elizalde-Peña, E.A.; Pojman, J.A.; Monte, F.D.; Luna-Bárcenas, G. Deep eutectic solvents as both active fillers and monomers for frontal polymerization. *J. Polym. Sci. Part A Polym. Chem.* **2013**, *51*, 1767–1773. [CrossRef]
62. Li, R.; Chen, G.; He, M.; Tian, J.; Su, B. Patternable transparent and conductive elastomers towards flexible tactile/strain sensors. *J. Mater. Chem. C* **2017**, *5*, 8475–8481. [CrossRef]
63. Li, R.; Zhang, K.; Chen, G.; Su, B.; Tian, J.; He, M.; Lu, F. Green polymerizable deep eutectic solvent (PDES) type conductive paper for origami 3D circuits. *Chem. Commun.* **2018**, *54*, 2304–2307.
64. Wang, M.; Li, R.; Chen, G.; Zhou, S.; Feng, X.; Chen, Y.; He, M.; Liu, D.; Song, T.; Qi, H. Highly Stretchable, Transparent, and Conductive Wood Fabricated by in Situ Photopolymerization with Polymerizable Deep Eutectic Solvents. *ACS Appl. Mater. Interfaces* **2019**, *11*, 14313–14321. [CrossRef] [PubMed]
65. Choi, H.-M.; Cho, J.Y. Microwave-mediated rapid tailoring of PET fabric surface by using environmentally-benign, biodegradable Urea-Choline chloride Deep eutectic solvent. *Fibers Polym.* **2016**, *17*, 847–856. [CrossRef]
66. Cho, J.Y.; Choi, H.-M.; Oh, K.W. Rapid hydrophilic modification of poly (ethylene terephthalate) surface by using deep eutectic solvent and microwave irradiation. *Text. Res. J.* **2016**, *86*, 1318–1327. [CrossRef]
67. Choi, S.; Choi, H.-M. Eco-friendly, expeditious depolymerization of PET in the blend fabrics by using a bio-based deep eutectic solvent under microwave irradiation for composition identification. *Fibers Polym.* **2019**, *20*, 752–759. [CrossRef]
68. Sert, E.; Yılmaz, E.; Atalay, F.S. Chemical recycling of polyethlylene terephthalate by glycolysis using deep eutectic solvents. *J. Polym. Environ.* **2019**, *27*, 2956–2962. [CrossRef]
69. Liu, B.; Fu, W.; Lu, X.; Zhou, Q. Zhang, S. Lewis acid-base synergistic catalysis for polyethylene terephthalate degradation by 1,3-dimethylurea/Zn (OAc)$_2$ deep eutectic solvent. *ACS Sustain. Chem. Eng.* **2019**, *7*, 3292–3300. [CrossRef]
70. Wang, Q.; Yao, X.; Geng, Y.; Zhou, Q.; Lu, X.; Zhang, S. Deep eutectic solvents as highly active catalysts for the fast and mild glycolysis of poly(ethylene terephthalate) (PET). *Green Chem.* **2015**, *17*, 2473–2479. [CrossRef]
71. Zhou, L.; Lu, X.; Ju, Z.; Liu, B.; Yao, H.; Xu, J.; Zhou, Q.; Hu, Y.; Zhang, S. Alcoholysis of polyethylene terephthalate to produce dioctyl terephthalate using choline chloride-based deep eutectic solvents as efficient catalysts. *Green Chem.* **2019**, *21*, 897–906. [CrossRef]

72. Musale, R.M.; Shukla, S.R. Deep eutectic solvent as effective catalyst for aminolysis of polyethylene terephthalate (PET) waste. *Int. J. Plast. Technol.* **2016**, *20*, 106–120. [CrossRef]
73. García-Argüelles, S.; García, C.; Serrano, M.C.; Gutiérrez, M.C.; Ferrer, M.L.; del Monte, F. Near-to-eutectic mixtures as bifunctional catalysts in the low-temperature-ring-opening-polymerization of ε-caprolactone. *Green Chem.* **2015**, *17*, 3632–3643. [CrossRef]
74. Pradeepkumar, P.; Rajendran, N.K.; Alarfaj, A.A.; Munusamy, M.A.; Rajan, M. Deep eutectic solvent-mediated FA-g-β-alanine-co-PCL drug carrier for sustainable and site-specific drug delivery. *ACS Appl. Bio. Mater.* **2018**, *1*, 2094–2109. [CrossRef]
75. Garcia-Arguelles, S.; Serrano, C.M.; Gutierrez, M.C.; Ferrer, L.M.; Yuste, L.; Rojo, F.; del Monte, F. Deep eutectic solvent-assisted synthesis of biodegradable polyesters with antibacterial properties. *Langmuir* **2013**, *29*, 9525–9534. [CrossRef] [PubMed]
76. Wang, J.; Han, J.; Khan, M.Y.; He, D.; Peng, H.; Chen, D.; Xie, X.; Xue, Z. Deep eutectic solvents for green and efficient iron-mediated ligand-free atom transfer radical polymerization. *Polym. Chem.* **2017**, *8*, 1616–1627. [CrossRef]
77. Wang, J.; Dong, X.; Yu, Q.; Baker, S.N.; Li, H.; Larm, N.E.; Baker, G.A.; Chen, L.; Tan, J.; Chen, M. Incorporation of antibacterial agent derived deep eutectic solvent into an active dental composite. *Dent. Mater.* **2017**, *33*, 1445–1455. [CrossRef] [PubMed]
78. Fu, N.; Liu, X.; Li, L.; Tang, B.; Row, K.H. Ternary choline chloride/caffeic acid/ethylene glycol deep eutectic solvent as both a monomer and template in a molecularly imprinted polymer. *J. Sep. Sci.* **2017**, *40*, 2286–2291. [CrossRef]
79. Li, G.; Tang, W.; Cao, W.; Wang, Q.; Zhu, T. Molecularly imprinted polymers combination with deep eutectic solvents for solid-phase extraction of caffeic acid from hawthorn. *Chin. J. Chromatogr.* **2015**, *33*, 792–798. [CrossRef]
80. Tang, W.; Row, K.H. Fabrication of water-compatible molecularly imprinted resin in a hydrophilic deep eutectic solvent for the determination and purification of quinolones in wastewaters. *Polymers* **2019**, *11*, 871. [CrossRef]
81. Liang, S.; Yan, H.; Cao, J.; Han, Y.; Shen, S.; Bai, L. Molecularly imprinted phloroglucinol-formaldehyde-melamine resin prepared in a deep eutectic solvent for selective recognition of clorprenaline and bambuterol in urine. *Anal. Chim. Acta* **2017**, *951*, 68–77. [CrossRef]
82. Xu, K.; Wang, Y.; Wei, X.; Chen, J.; Xu, P.; Zhou, Y. Preparation of magnetic molecularly imprinted polymers based on a deep eutectic solvent as the functional monomer for specific recognition of lysozyme. *Mikrochim. Acta* **2018**, *185*, 146. [CrossRef] [PubMed]
83. Jin, F.-L.; Li, X.; Park, S.-J. Synthesis and application of epoxy resins: A review. *J. Ind. Eng. Chem.* **2015**, *29*, 1–11. [CrossRef]
84. Chruściel, J.J.; Leśniak, E. Modification of epoxy resins with functional silanes, polysiloxanes, silsesquioxanes, silica and silicates. *Prog. Polym. Sci.* **2015**, *41*, 67–121. [CrossRef]
85. Ma, Y.; Liu, H.; Wu, J.; Yuan, L.; Wang, Y.; Du, X.; Wang, R.; Marwa, P.W.; Petlulu, P. The adverse health effects of bisphenol A and related toxicity mechanisms. *Environ. Res.* **2019**, *176*, 108575. [CrossRef]
86. Chen, X.; Hou, J.; Gu, Q.; Wang, Q.; Gao, J.; Sun, J.; Fang, Q. A non-bisphenol-A epoxy resin with high Tg derived from the bio-based protocatechuic Acid:Synthesis and properties. *Polymer* **2020**, *195*, 122443. [CrossRef]
87. Hsu, Y.-I.; Huang, L.; Asoh, T.-A.; Uyama, H. Anhydride-cured epoxy resin reinforcing with citric acid-modified cellulose. *Polym. Degrad. Stab.* **2020**, *178*, 109213. [CrossRef]
88. Lu, J.; Yan, F.; Texter, J. Advanced applications of ionic liquids in polymer science. *Prog. Polym. Sci.* **2009**, *34*, 431–448. [CrossRef]
89. Carvalho, A.P.A.; Santos, D.F.; Soares, B.G. Epoxy/imidazolium-based ionic liquid systems: The effect of the hardener on the curing behavior, thermal stability, and microwave absorbing properties. *J. Appl. Polym. Sci.* **2020**, *137*, 48326. [CrossRef]
90. Nguyen, T.K.L.; Livi, S.; Soares, B.G.; Pruvost, S.; Duchet-Rumeau, J.; Gérard, J.-F. Ionic liquids: A new route for the design of epoxy networks. *ACS Sustain. Chem. Eng.* **2015**, *4*, 481–490. [CrossRef]
91. Sidorov, O.I.; Vygodskii, Y.S.; Lozinskaya, E.I.; Milekhin, Y.M.; Matveev, A.A.; Poisova, T.P.; Ferapontov, F.V.; Sokolov, V.V. Ionic liquids as curing catalysts for epoxide-containing compositions. *Polym. Sci. Ser. D* **2017**, *10*, 134–142. [CrossRef]

92. Donato, R.K.; Matějka, L.; Schrekker, H.S.; Pleštil, J.; Jigounov, A.; Brus, J.; Šlouf, M. The multifunctional role of ionic liquids in the formation of epoxy-silica nanocomposites. *J. Mater. Chem.* **2011**, *21*, 13801–13810. [CrossRef]
93. Doolan, P.C.; Gore, P.H.; Hollingworth, R.H.; Waters, D.N.; Al-Ka'Bi, J.F.; Farooqi, J.A. Kinetic studies of lewis acidity. Part 2. Catalysis by tin (IV) chloride, by some organotin (IV) chlorides, and by tin (II) chloride of the anionotropic rearrangement of henylpropnl in tetramethylene sulphone solution. *J. Chem. Soc.* **1986**, *17*, 501–506. [CrossRef]
94. Mąka, H.; Spychaj, T.; Sikorski, W. Deep eutectic ionic liquids as epoxy resin curing agents. *Int. J. Polym. Anal. Charact.* **2014**, *19*, 682–692. [CrossRef]
95. Lionetto, F.; Timo, A.; Frigione, M. Curing kinetics of epoxy-deep eutectic solvent mixtures. *Thermochim. Acta* **2015**, *612*, 70–78. [CrossRef]
96. Chen, X.; Xiong, L. Li, H.; Chen, X. Research progress of deep eutectic solvent in lignocellulose pretreatment to promote enzymatic hydrolysis efficiency. *Adv. New Renew. Energy* **2019**, *7*, 415–422.
97. Xie, Y.; Guo, X.; Lv, Y.; Wang, H. Research progress of deep eutectic solvent in the dissolution and separation of fiber raw materials. *Chem. Ind. For. Prod.* **2019**, *39*, 11–18.
98. Sun, X.; Ling, C.; Lian, H. The pyrolysis and pyrolysis kinetics of phenol-formaldehyde resin modified with lignin in deep eutectic solvent. *China For. Prod. Ind.* **2016**, *43*, 19–24.
99. Muylaert, I.; Verberckmoes, A.; De Decker, J.; Van Der Voort, P. Ordered mesoporous phenolic resins: Highly versatile and ultra stable support materials. *Adv. Colloid Interface Sci.* **2012**, *175*, 39–51. [CrossRef]
100. Effendi, A.; Gerhauser, H.; Bridgwater, A.V. Production of renewable phenolic resins by thermochemical conversion of biomass: A review. *Renew. Sustain. Energy Rev.* **2008**, *12*, 2092–2116. [CrossRef]
101. Wang, B.; Wang, Y.-P.; Zhou, P. Liu, Z.-Q.; Luo, S.-Z.; Chu, W.; Guo, Z. Formation of poly (acrylic acid)/alumina composite via in situ polymerization of acrylic acid adsorbed within oxide pores. *Colloids Surf. A Physicochem Eng. Asp.* **2017**, *514*, 168–177. [CrossRef]
102. Patil, Y.; Ameduri, B. Advances in the (co)polymerization of alkyl 2-trifluoromethacrylates and 2-(trifluoromethyl) acrylic acid. *Prog. Polym. Sci.* **2013**, *38*, 703–739. [CrossRef]
103. Ballard, N.; Asua, J.M. Radical polymerization of acrylic monomers: An overview. *Prog. Polym. Sci.* **2018**, *79*, 40–60. [CrossRef]
104. Sánchez-Leija, R.J.; Pojman, J.A.; Luna-Bárcenas, G.; Mota-Morales, J.D. Controlled release of lidocaine hydrochloride from polymerized drug-based deep-eutectic solvents. *J. Mater. Chem. B* **2014**, *2*, 7495–7501. [CrossRef] [PubMed]
105. Tomé, L.I.N.; Baião, V.; da Silva, W.; Brett, C.M.A. Deep eutectic solvents for the production and application of new materials. *Appl. Mater. Today* **2018**, *10*, 30–50. [CrossRef]
106. Pang, K.; Kotek, R.; Tonelli, A.E. Review of conventional and novel polymerization processes for polyesters. *Prog. Polym. Sci.* **2006**, *31*, 1009–1037. [CrossRef]
107. Zhang, X. Hyperbranched aromatic polyesters: From synthesis to applications. *Prog. Org. Coat.* **2010**, *69*, 295–309. [CrossRef]
108. Gao, Y.; Romero, P.; Zhang, H.; Huang, M.; Lai, F. Unsaturated polyester resin concrete: A review. *Constr. Build. Mater.* **2019**, *228*, 116709. [CrossRef]
109. Liu, Y.; Yao, X.; Yao, H.; Zhou, Q.; Xin, J.; Lu, X.; Zhang, S. Degradation of poly (ethylene terephthalate) catalyzed by metal-free choline-based ionic liquids. *Green Chem.* **2020**, *22*, 3122–3131. [CrossRef]
110. Zhou, X.D.; Chen, K.; Guo, J.L.; Zhang, Y.F. Synthesis and application of quaternary ammonium-functionalized hyperbranched polyester Antibacterial Agent. *Adv. Mater. Res.* **2013**, *796*, 412–415. [CrossRef]
111. Bures, F. Quaternary ammonium compounds: Simple in structure, complex in application. *Top. Curr. Chem.* **2019**, *377*, 14. [CrossRef]
112. Vasapollo, G.; Sole, R.D.; Mergola, L.; Lazzoi, M.R.; Scardino, A.; Scorrano, S.; Mele, G. Molecularly imprinted polymers: Present and future prospective. *Int. J. Mol. Sci.* **2011**, *12*, 5908–5945. [CrossRef] [PubMed]
113. Turiel, E.; Esteban, A.M. Molecularly imprinted polymers. *Solid Phase Extr.* **2020**, 215–233.
114. BelBruno, J.J. Molecularly imprinted polymers. *Chem. Rev.* **2019**, *119*, 94–119. [CrossRef] [PubMed]

115. Kartal, F.; Cimen, D.; Bereli, N.; Denizli, A. Molecularly imprinted polymer based quartz crystal microbalance sensor for the clinical detection of insulin. *Mater. Sci. Eng.* **2019**, *97*, 730–737. [CrossRef]
116. Sibrian-Vazquez, M.; Spivak, D.A. Characterization of molecularly imprinted polymers employing crosslinkers with nonsymmetric polymerizable groups. *J. Polym. Sci. Part A Polym. Chem.* **2004**, *42*, 3668–3675. [CrossRef]

© 2020 by the authors. Licensee MDPI, Basel, Switzerland. This article is an open access article distributed under the terms and conditions of the Creative Commons Attribution (CC BY) license (http://creativecommons.org/licenses/by/4.0/).

Review

Ionic Liquids/Deep Eutectic Solvents-Based Hybrid Solvents for CO_2 Capture

Yanrong Liu [1,2], Zhengxing Dai [3], Fei Dai [4,5,*] and Xiaoyan Ji [1,*]

1. Energy Engineering, Division of Energy Science, Luleå University of Technology, 97187 Luleå, Sweden; yanrong.liu@ltu.se
2. Swerim AB, Box 812, SE-97125 Luleå, Sweden
3. State Key Laboratory of Material-Oriented Chemical Engineering, Nanjing Tech University, Nanjing 211816, China; 201861104285@njtech.edu.cn
4. CAS Key Laboratory of Green Process and Engineering, Beijing Key Laboratory of Ionic Liquids Clean Process, State Key Laboratory of Multiphase Complex Systems, Institute of Process Engineering, Chinese Academy of Sciences, Beijing 100190, China
5. School of Chemical Engineering, University of Chinese Academy of Sciences, Beijing 100049, China
* Correspondence: fdai@ipe.ac.cn (F.D.); xiaoyan.ji@ltu.se (X.J.)

Received: 11 October 2020; Accepted: 26 October 2020; Published: 29 October 2020

Abstract: The CO_2 solubilities (including CO_2 Henry's constants) and viscosities in ionic liquids (ILs)/deep eutectic solvents (DESs)-based hybrid solvents were comprehensively collected and summarized. The literature survey results of CO_2 solubility illustrated that the addition of hybrid solvents to ILs/DESs can significantly enhance the CO_2 solubility, and some of the ILs-based hybrid solvents are super to DESs-based hybrid solvents. The best hybrid solvents of IL–H_2O, IL–organic, IL–amine, DES–H_2O, and DES–organic are [DMAPAH][Formate] (2.5:1) + H_2O (20 wt %) (4.61 mol/kg, 298 K, 0.1 MPa), [P_{4444}][Pro] + PEG400 (70 wt %) (1.61 mol/kg, 333.15 K, 1.68 MPa), [DMAPAH][Formate] (2.0:1) + MEA (30 wt %) (6.24 mol/kg, 298 K, 0.1 MPa), [TEMA][Cl]-GLY-H_2O 1:2:0.11 (0.66 mol/kg, 298 K, 1.74 MPa), and [Ch][Cl]-MEA 1:2 + DBN 1:1 (5.11 mol/kg, 298 K, 0.1 MPa), respectively. All of these best candidates show higher CO_2 solubility than their used pure ILs or DESs, evidencing that IL/DES-based hybrid solvents are remarkable for CO_2 capture. For the summarized viscosity results, the presence of hybrid solvents in ILs and DESs can decrease their viscosities. The lowest viscosities acquired in this work for IL–H_2O, IL–amine, DES–H_2O, and DES–organic hybrid solvents are [DEA][Bu] + H_2O (98.78 mol%) (0.59 mPa·s, 343.15 K), [BMIM][BF_4] + DETA (94.9 mol%) (2.68 mPa·s, 333.15 K) [L-Arg]-GLY 1:6 + H_2O (60 wt %) (2.7 mPa·s, 353.15 K), and [MTPP][Br]-LEV-Ac 1:3:0.03 (16.16 mPa·s, 333.15 K) at 0.1 MPa, respectively.

Keywords: ionic liquid; deep eutectic solvents; hybrid solvent; CO_2 solubility; Henry's constant; viscosity

1. Introduction

CO_2 emission is an urgent issue due to its main contribution to global warming [1]. It has been evidenced that CO_2 capture is a promising route to mitigate CO_2 emissions, and in general, cost, energy demand, and environmental impact need to be considered when for selecting the potential CO_2 capture technologies [2]. At present, the absorption technology with 30 wt % MEA aqueous solution is the commercialized one. However, this technology with the corresponding solvent has the drawbacks of high energy demand (4.2 GJ/t CO_2), high cost ($0.19–1.31/t CO_2), low thermal and chemical stability, and high volatility and corrosion [3–6], which underlines the necessity for developing greener and more efficient solvent for CO_2 capture.

Compared with the traditional amine-based solvent, the emerging new absorption solvents of ionic liquids (ILs) and deep eutectic solvents (DESs) have attracted more and more attention due to the merits of recyclability, good solvent stability, low energy demand, and environmentally friendly nature [7]. However, some of the ILs (e.g., [P$_{4442}$][Cy-Suc], 2567 mPa·s at 303.15 K) [8] and DESs (e.g., [MTPP][Br]-GLY 1:4, 1658 mPa·s at 298.15 K) [9] have high viscosities that influence the rate of absorptions, inhibiting their industry applications. To cope with this disadvantage, hybrid ILs or DESs with water have been strongly recommended [9–14]. For example, Zhang et al. investigated the mass transfer feature in [BMIM][NO$_3$] + H$_2$O, evidencing that mass transfer increases with the increase of water content, e.g., the mass transfer of [BMIM][NO$_3$] (95 wt %) + H$_2$O (5 wt %) is 0.55×10^5 m·s^{-1}, while it is 0.64×10^5 m·s^{-1} for [BMIM][NO$_3$] (90 wt %) + H$_2$O (10 wt %), which may due to the decrease of viscosity that from 35.74 ([BMIM][NO$_3$] (95 wt %) + H$_2$O (5 wt %)) to 12.65 mPa·s ([BMIM][NO$_3$] (90 wt %) + H$_2$O (10 wt %)) [13]. Sarmad et al. studied the viscosity of DES, finding that a small amount of water has a significant effect on DES viscosity [9]. For instance, at 298 K, the viscosity of [TEMA][Cl]-GLY (1:2) is 236.59 mPa·s, while it is 72.75 mPa·s for [TEMA][Cl]-GLY-H$_2$O (1:2:0.055).

Except for adding water to decrease the viscosity, organic solvents (e.g., PEG, TEG, and TG) can substitute water totally or partially to decrease the viscosity or to overcome high energy demand in IL + H$_2$O [15–17]. The experimental result indicates that TG can significantly decrease the viscosity of [P$_{66614}$][4-Triz], especially at low temperatures, i.e., the viscosity of [P$_{66614}$][4-Triz] is 4640 mPa·s at 278.15 K, while it is 163 mPa·s for [P$_{66614}$][4-Triz] + TG (58.2 mol%) [18]. Liu et al. found that the addition of a certain amount of PEG200 to [Cho][Gly] + H$_2$O not only decreases the viscosity but also enhances the CO$_2$ solubility, as well as decreases the desorption enthalpy [17].

To further take the benefits of both neoteric and conventional solvents, IL–amine-based and superbase–amine-based DES hybrid absorbents have also been proposed and developed [19–25]. These hybrid solvents possess certain advantages of low energy demand, low water evaporation, and high CO$_2$ solubility compared to the commercialized MEA aqueous solution [26]. For example, Yang et al. reported that [BMIM][BF$_4$] (40 wt %) + MEA (30 wt %) + H$_2$O (30 wt %) has higher CO$_2$ solubility than MEA (30 wt %) + H$_2$O (70 wt %), and the energy demand reduced by 37.2% with respect to MEA (30 wt %) + H$_2$O (70 wt %) [21]. The CO$_2$ solubilities of four functionalized ILs in MDEA aqueous solution were investigated and compared with MEA + MDEA aqueous solution [27], showing that the CO$_2$ solubility of [N$_{2222}$][Lys] (15 wt %) + MDEA (15 wt %) + H$_2$O (70 wt %) (0.74 mol/mol) > [N$_{1111}$][Lys] (15 wt %) + MDEA (15 wt %) + H$_2$O (70 wt %) (0.69 mol/mol) > [N$_{2222}$][Gly] (15 wt %) + MDEA (15 wt %) + H$_2$O (70 wt %) (0.64 mol/mol) > [N$_{1111}$][Gly] (15 wt %) + MDEA (15 wt %) + H$_2$O (70 wt %) (0.56 mol/mol) > MEA (15 wt %) + MDEA (15 wt %) + H$_2$O (70 wt %) (0.36 mol/mol). For DES hybrid solvent of [Ch][Cl]-MEA 1:2 + DBN with volume ratio of 1:1, its CO$_2$ solubility improved from 3.29 to 5.11 mol/kg compared with [Ch][Cl]-MEA 1:2 at 298.15 K [25].

To develop the potential IL/DES-based hybrid solvents for CO$_2$ capture, CO$_2$ solubility (in accordance with Henry's constant for physical absorption) and viscosity are two key properties. Furthermore, the selectively for physical absorption and the CO$_2$ absorption enthalpy for chemical absorption are other concerns in development, while the research is still limited, especially when compared to those for CO$_2$ solubility and viscosity. Several reviews for IL/DES-based hybrid solvents from the aspect of CO$_2$ solubility, Henry's constant, and viscosity have been published. Babamohammadi et al. summarized the viscosities of IL + H$_2$O and IL + MEA/EG + H$_2$O until 2014 and the CO$_2$ solubilities of ILs (imidazolium- and ammonium-based ILs)-amine hybrid solvents since 2008 [28]. Gao et al. summarized the CO$_2$ solubility of 18 kinds of IL-amine based hybrid solvent until 2015. Huang et al. reviewed the advantages and disadvantages of five kinds of IL–hybrid solvents (i.e., IL–organic, normal IL–amine, normal IL aqueous–amine, functionalized IL–amine, and functionalized IL aqueous–amine) until 2016 for CO$_2$ capture [26], finding that IL–hybrid solvents can significantly reduce the viscosity. Lian et al. introduced the ILs–hybrid processes for CO$_2$ capture and compared the CO$_2$ solubilities for IL–DEA/DMEE/ethanol, indicating that IL–ethanol has the highest solubility of 2.3 mol/mol [29]. Recently, more IL-based hybrid solvents have been developed combined with property measurements, making it necessary to update the latest research progress. Meanwhile, to the best of our knowledge, there is no review article for the DES-based hybrid solvents.

To fulfill this gap and to promote the technology development on CO_2 capture in IL/DES-based hybrid solvents, this review summarizes the CO_2 solubilities (including Henry's constants) and viscosities of IL-based hybrid solvents since 2016 and DES-based hybrid solvents since 2014 to avoid the repetition of the published reviews. Finally, the best candidates for IL/DES-based hybrid solvents were obtained and compared with each other.

2. ILs-Based Hybrid Solvents

Regarding the CO_2 solubilities for 73 kinds of IL-H_2O and 37 kind of IL-organic-based hybrid solvents since 2016, 28 types of IL-amine hybrid solvents since 2018 together with 62 Henry's constants have been reported. The results were collected and summarized in Tables 1 and 2. The viscosities for 30 kinds of IL-H_2O and 121 IL-organic/organic aqueous solution hybrid solvents since 2016, and 15 kinds of IL-amine/amine aqueous solution hybrid solvents since 2018 have been determined, and these are listed in Table 3. The full names of ILs-based hybrid solvents are displayed in Table S1.

2.1. CO_2 Solubility and Henry's Constant

2.1.1. IL-H_2O

The effect of the addition of H_2O in [DMAPAH][Formate] (0.5:1, 1.0:1, 2.0:1, 2.5:1) was studied by Vijayaraghavan et al. [30]. Except for the molar ratio of amine to acid of 0.5:1, CO_2 solubility initially increased up to a certain water amount of 20 wt % (Figure 1); then, it steadily decreased as the H_2O concentration increased. As shown in Figure 1, the best candidate for CO_2 capture is [DMAPAH][Formate] (2.5:1) + H_2O (20 wt %) (5.69 mol/kg). The CO_2 solubility in [TMGH][Im] + H_2O (1–25 wt %) was studied by Li et al. [12]. The result indicates that CO_2 solubility first increased at a water content range of 1–7 wt %, and then, it decreased when the water content was larger than 7 wt % in [TMGH][Im], resulting in the best absorption capacity of 4.23 mol/kg for [TMGH][Im] + H_2O (7 wt %). Huang et al. reported that a [P_{4442}][Suc] structure with basic anion can improve the reaction of CO_2 and [P_{4442}][Suc] aqueous solution by forming bicarbonate and conjugate acid [31]. The result indicates that the CO_2 solubility of [P_{4442}][Suc] + H_2O (3.3 wt %) (1.9 mol/mol) is slightly higher than [P_{4442}][Suc] (1.87 mol/mol); however, the addition of 8.8 and 17.6 wt % of H_2O to [P_{4442}][Suc] has a negative effect on their CO_2 solubilities compared with [P_{4442}][Suc]. As shown in Table 1, various compositions of [P_{4444}][HCOO] in water were measured at 0.1 MPa in the temperature range of 248.75–324.65 K for CO_2 solubility and compared with [N_{2224}][Ac] and [N_{2222}][Ac] [32]. The results indicate that the CO_2 solubility of [P_{4444}][HCOO] + H_2O first increased from 0.01 to 1 mol/mol at the water contents range of 29–66 mol% and then decreased with the increase of water contents of 70–91 mol%. Temperature also affects the absorption performance of [P_{4444}][HCOO] + H_2O. For example, at 273.15 K, the CO_2 solubility of [P_{4444}][HCOO] + H_2O (<80 mol%) is higher than [N_{2224}][Ac] + H_2O and [N_{2222}][Ac] + H_2O, while at 298.15 K, it is higher than [N_{2224}][Ac] + H_2O at a water content less than 50 mol%, but it is lower at a water content high than 50 mol%. For 323.15 K, the CO_2 solubility in [P_{4444}][HCOO] + H_2O (<80 mol%) is lower than [N_{2222}][Ac] + H_2O. Nathalia et al. [33] evidenced that there are three processes for CO_2 capture in [BMMIM][Im] and [BMMIM][Ac] aqueous solutions, i.e., physical, CO_2–imidazolium adduct generation, and bicarbonate formation, resulting in a maximum CO_2 solubility of 8.15 mol/mol in [BMMIM][Im] + H_2O (99.9 mol%). The CO_2 solubility in [P_{4443}][Gly] + H_2O (59.9, 80.1, 90, 95 wt %) was measured at temperatures ranging from 278.14 to 348.05 K and pressures of 0.1–7.75 MPa [34]. A feature of physical absorption in these hybrid solvents was observed. The best CO_2 solubility of 2.44 mol/kg was acquired for [P_{4443}][Gly] + H_2O (59.9 wt %) at 298.06 K and 4.6 MPa. Aghaie et al. tested the CO_2 solubilities of [HMIM][Tf_2N], [HMIM][FAP], and [BMIM][Ac] aqueous solutions [35]. The result indicates that the CO_2 solubility reduced by 45% in these three ILs aqueous solutions compared to the solubility of CO_2 in pure IL at 298 K and water content of 10 wt %.

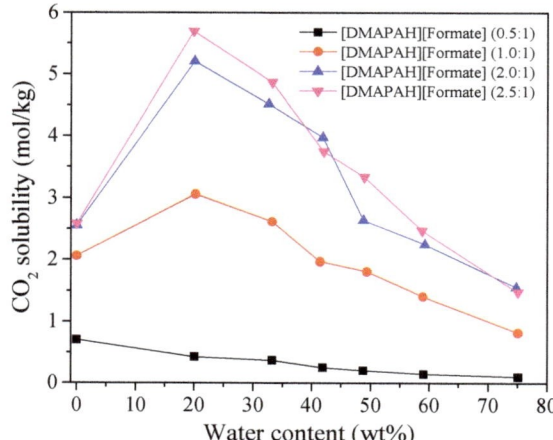

Figure 1. Effects of water on CO_2 solubility in [DMAPAH][Formate]. The values in Figure 1 are cited from Vijayaraghavan et al. [30]. Copyright 2018 Elsevier.

2.1.2. IL–Organic/Organic Aqueous Solution

Huang et al. investigated the CO_2 capture capacity in [TETAH][Lys] + ethanol + H_2O [36], finding that the CO_2 solubility first increases with the increase of volume ratio of ethanol in water from 10:0 to 5:5 v/v, and then, it decreases from 5:5 to 2:8 v/v. Compared with the reported results, the maximum CO_2 solubility of 2.45 mol/mol for [TETAH][Lys] + ethanol + H_2O (5:5 v/v) is higher than those of [P_{6444}][Lys] [37], [C_2NH_2MIM][Lys] [38], and [TETAH][Lys] + H_2O. Taheri et al. measured the CO_2 solubility of [AMIM][Tf_2N] + methanol at temperature and pressure ranges of 313.2–353.2 K and 0.98–6.19 MPa, respectively, indicating that the presence of methanol in [AMIM][Tf_2N] enhances the CO_2 solubility and results in a maximum of 3.89 mol/mol [39].

In order to overcome the drawbacks of high CO_2 capture enthalpy and water volatilization in ILs aqueous solution, PEG200 was introduced in [Cho][Gly] + H_2O [17]. CO_2 solubility was measured in such solvent at 308.15–338.15 K and pressure lower than 0.68 MPa, and the CO_2 desorption enthalpy was estimated. Owing to its physical–chemical properties, [Cho][Gly] (30 wt %) + PEG200 (30 wt %) + H_2O (40 wt %) has a higher CO_2 solubility (0.41–1.23 mol/kg) and regeneration efficiency (95%) compared with [Cho][Gly] (30 wt %) + H_2O (70 wt %). Li et al. found that the addition of PEG200 to [Cho][Pro] not only improves the absorption rate but also enhances the desorption efficient, resulting in a maximum CO_2 solubility of about 0.6 mol/mol for [Cho][Pro] + PEG200 with the mass ratios of 1:1, 1:2, 1:3, respectively [40]. PEG400 was introduced to [P_{4444}][Gly], [P_{4444}][Ala], and [P_{4444}][Pro] [41], evidencing that [P_{4444}][Pro] + PEG400 (70 wt %) (1.61 mol/kg) > [P_{4444}][Gly] + PEG400 (70 wt %) (1.58 mol/kg) > [P_{4444}][Ala] + PEG400 (70 wt %) (1.57 mol/kg). The effect of three types of PEG (i.e., PEG200, PEG300, and PEG400) and water content on the CO_2 solubility in [DETAH][Br] and [DETAH][BF_4] at 293.15 K and 0.1 MPa was investigated by Chen et al. [42]. The result evidenced that the CO_2 solubility follows the order of [DETAH][Br] + PEG200 (1.18 mol/mol) > [DETAH][Br] + PEG300 (0.87 mol/mol) > [DETAH][BF_4] + PEG200 (0.65 mol/mol) > [DETAH][Br] + PEG300 (0.32 mol/mol) at a mass ratio of [DETAH][Br]:PEG = 1:4. Additionally, the CO_2 solubility in [DETAH][Br] + PEG200 + H_2O (4.7 wt %) (1.18 mol/mol) is higher than that in [DETAH][Br] + PEG200 + H_2O (1.3 wt %) (1.05 mol/mol), which may be because water weakens the interaction of the IL cation and anion and enhances the interaction with CO_2. Due to the high CO_2 solubility of PEO1000 (0.35 mol/mol, 323 K, 4.98 MPa), it is introduced to [N_{4111}][Tf_2N]. A maximum CO_2 solubility of 1.16 mol/mol was acquired at 323 K, 4.99 MPa for [N_{4111}][Tf_2N] + PEO1000 (75 mol%), which is higher than that of pure [N$_{4111}$][Tf$_2$N] (0.14 mol/mol, 323 K, 5 MPa) [43]. Additionally, a higher amount of PEO1000 in [N$_{4111}$][Tf$_2$N] corresponds to higher CO$_2$ solubility, which is attributed to the strong interaction between CO$_2$ and PEO. However, Jiang et al. [44] found that increasing the molar fraction of TEG in [BMIM][BF$_4$]/[BMIM][BF$_4$] + H$_2$O results in a decrease of CO$_2$ solubilities, which is on the contrary of the result from Lepre et al. [43]. Moreover, with increasing the [BMIM][BF$_4$] contents in [BMIM][BF$_4$] + TEG mixtures, the Henry's constant is increased (Table 2) [44]. The Henry's constant of [BMIM][BF$_4$] + TEG is higher than [BMIM][BF$_4$] but lower than TEG, indicating that CO$_2$ is more soluble in [BMIM][BF$_4$]. The CO$_2$ solubilities of [P$_{66614}$][3-Triz] + TG (30 mol%) and [P$_{66614}$][4-Triz] + TG (30 mol%) at 313.15–353.6 K and pressure less than 3 MPa were measured by Fillion et al. [18]. The CO$_2$ solubility of [P$_{66614}$][4-Triz] + TG (30 mol%) is 2.23 mol/mol at 313.15 K and 3.03 MPa, which is higher than [P$_{66614}$][3-Triz] + TG (30 mol%) (1.55 mol/mol, 313.15 K, and 2.68 MPa). [TEPAH][2-MI] combined with propan-1-ol (NPA) and EG was used for CO$_2$ capture [45]. The result indicates that the CO$_2$ solubility in [TEPAH][2-MI] + NPA + EG can reach up to 1.72 mol/mol, which was much higher than that of [C$_3$OHmim]Cl + MEA (0.3 mol/mol) [46], AMP + MEA – H$_2$O (0.5 mol/mol) [47], [P$_{66614}$][Gly] (1.26 mol/mol) [48], and TETA + AMP + ethanol (1.03 mol/mol) [49].

2.1.3. IL–Amine/Amine Aqueous Solution

Three base-rich diamino ILs of [DMAPAH][Formate], [DMEDAH][Formate], and [DMAPAH][Octanoate] were synthesized with different molar ratios of base to acid (0.5:1, 1.0:1, 2.0:1, and 2.5:1, respectively) and hybrid with MEA for CO$_2$ capture, respectively [30]. According to Table 1 and Figure 2, the hybrid solvents of the synthesized ILs with an additional MEA showed enhanced CO$_2$ solubility, which agrees with the studies from Zeng et al. [50] and Meng et al. [51] that applied MDEA and DMEE as the hybrid solvents to [DMAPAH][Ac] and [N$_{1111}$][Lys], respectively. Among them, [DMAPAH][Formate] (2.0:1) + MEA (30 wt%) with 6.24 mol/kg was identified to be the best one for CO$_2$ capture. The CO$_2$ capture performance in [BMPyrr][DCA] (5 wt%) + DEA (35 wt%) + H$_2$O (60 wt%) and [BMPyrr][DCA] (10 wt%) + DEA (30 wt%) + H$_2$O (60 wt%) were studied by Salleh et al. [52] and compared with DEA (40 wt%) + H$_2$O (60 wt%). The result indicates that the CO$_2$ solubility increases with increasing the [BMPyrr][DCA] amount in the hybrid solvent. However, the CO$_2$ solubilities of these two hybrid solvents are lower than those in DEA (40 wt%) + H$_2$O (60 wt%).

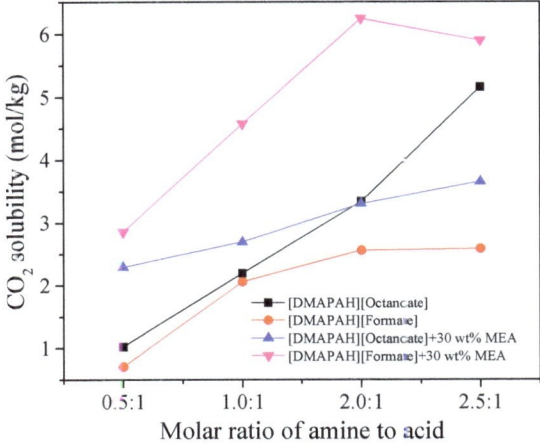

Figure 2. Effects of MEA on CO$_2$ solubilities in [DMAPAH][Formate] and [DMAPAH][Octanoate] [30]. Copyright 2018 Elsevier.

In conclusion, (1) a certain amount of water in ILs (mainly for chemical-based ILs) can enhance the CO_2 solubility, due to the decrease in viscosity and the formation of new products (e.g., carbamate and bicarbonate). However, excess water in ILs corresponds to a low ILs concentration and results in the decrease of CO_2 solubility; (2) the IL–organic and IL–organic aqueous solution as absorbents exhibit remarkable CO_2 capture performances, including high absorption capacity and low desorption enthalpy. The organic molecular weight, type, and water content in ILs can affect their CO_2 capture performance. Based on the summarized result, the organic solvent with low molecular weight together with a certain amount of water is beneficial for capturing CO_2; (3) IL–MEA shows better CO_2 capture performance than that of IL–MDEA and IL–DMEE; additionally, the IL–amine based hybrid solvent has higher CO_2 solubility than that of IL–H_2O and IL–organic hybrid solvents. The best for each of them are [DMAPAH][Formate] (2.0:1) + MEA (30 wt %) (6.24 mol/kg, 298 K, 0.1 MPa), [DMAPAH][Formate] (2.5:1) + H_2O (20 wt %) (4.61 mol/kg, 298 K, 0.1 MPa), and [P_{4444}][Pro] + PEG400 (70 wt %) (1.61 mol/kg, 333.15 K, 1.68 MPa). Sometimes, the presence of water in IL–organic/amine hybrid solvents improves the CO_2 solubility.

2.2. Viscosity

2.2.1. IL–H_2O

The viscosities of IL–H_2O hybrid solvents are given in Table 3. Figure 3 displays the viscosities of [DEA][Bu] + H_2O [14]; the result indicated that the [DEA][Bu] has a strong hygroscopic characteristic. The viscosity of [DEA][Bu] + H_2O decreases with the increase of water content and temperature. Yasaka et al. found that the viscosities of [P_{4444}][HCOO] + H_2O decreased with water contents from 25 (356 mPa·s) to 91 mol% (14.4 mPa·s), corresponding to an increase of the CO_2 solubility between the water conetnt of 25 and 50 mol%, and then, they decrease from 50 to 91 mol%, which is regarded as the typical property of carboxylate ILs [32]. Aghaie et al. measured the viscosities of [HMIM][Tf_2N], [HMIM][FAP], and [BMIM][Ac] aqueous solutions at 298–333 K, 2 MPa, and water mass percentages of 0.1, 1, 2, 5, and 10 wt %, respectively [35]. The result indicates that water has a significant effect on [BMIM][Ac] viscosity, e.g., the viscosity of [BMIM][Ac] decreased from 47.64 (pure IL) to 3.77 (10 wt % H_2O) mPa·s at 333 K. However, for [HMIM][Tf_2N] and [HMIM][FAP], their viscosities only slightly decrease at 0.1–10 wt % water. For example, the viscosity of [HMIM][FAP] at 333 K is 20.72 mPa·s, while it is 20.47 mPa·s for [HMIM][FAP] + H_2O (10 wt %). Additionally, increasing the water amount in these three ILs results in the decrease of CO_2 solubility.

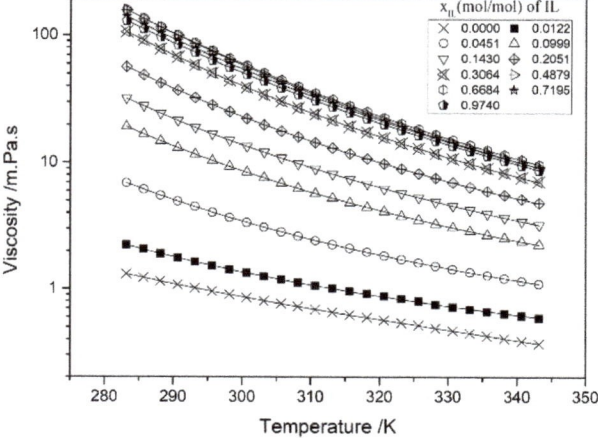

Figure 3. Viscosity of [DEA][Bu] + H_2O, 283.15–343.15 K, 0.1 MPa [14]. Copyright 2018 Elsevier.

2.2.2. IL–Organic/Organic Aqueous Solution

For [TETAH][Lys] + H_2O + ethanol, it was observed that its viscosity increased with the decrease amount of ethanol, corresponding to a decreased CO_2 solubility [36]. Liu et al. found that the viscosities of [Cho][Gly] (84.3–7.2 mPa·s), PEG200 (31.8–8.8 mPa·s), and [Cho][Gly] + PEG200 (70 wt %) (101.3–28.6 mPa·s) are much higher than that of [Cho][Gly]/H_2O + PEG200 (30 wt %) (7.96–3.43 mPa·s) at 308.15–338.15 K [17]. To avoid the decrease of absorption rate by increasing PEG200, less than 30 wt % of PEG200 was recommended to [Cho][Gly] + PEG200 + H_2O. The viscosities of [P_{4444}][Gly], [P_{4444}][Ala], and [P_{4444}][Pro] hybrid with PEG400 were measured at 298.15–393.15 K and 0.1 MPa [41]. The result indicates that the viscosities of these amino-acid ILs + PEG400 are about half with respect to the pure amino acid ILs at 298.15 K. Chen et al. reported that the viscosity of the [DETAH][Br] + PEG200 (80 wt %) is 71.7 mPa·s at 293.15 K and 0.1 MPa [42]. Despite the viscosities' increases with the increasing long-chain polymers, i e., [N_{4111}][Tf_2N] + PEO222 < [N_{4111}][Tf_2N] + PEO500 < [N_{4111}][Tf_2N] + PEO1000, [N_{4111}][Tf_2N] + PEO1000 was proposed for CO_2 capture due to the higher CO_2 solubility of the pure PEO1000 [43]. The viscosities for 18 kinds of ILs (imidazolium- and phosphonium-based) + TG at different of mole ratio of each of these ILs at 278.15–323.15 K and 0.1 MPa were studied [18], evidencing that the presence of TG can significantly decrease the viscosity, resulting in about 50 mPa·s for these hybrid solvents (Table 3).

2.2.3. IL–Amine

As shown in Table 3, the viscosities of the [N_{1111}][Lys] + DMEE and [BMIM][BF_4] + DETA hybrid solvents increased with increasing the content of ILs and decreasing temperature [51,53]. Based on the study of Meng et al. [51], the viscosity of [N_{1111}][Lys] + DMEE significantly decreased when the IL is <60 wt % compared to the pure [N_{1111}][Lys].

From Table 3, it can be found that the viscosities of IL-based hybrid solvents are very sensitive to H_2O, organic, and amine solvents. Their viscosities can significantly decrease with the increased amount of H_2O, PEG, TEG, TG, DMEE, and DETA; however, it increases with the increased amount of methanol. The lowest viscosity obtained from Table 3 for IL–H_2O, IL–organic, and IL–amine based hybrid solvents are [DEA][Bu] + H_2O (98.78 mol%) (0.59 mPa·s, 343.15 K), [N_{1114}][Tf_2N] + PEO222 (75.06 mol%) (2.57 mPa·s, 353.15 K), and [BMIM][BF_4] + DETA (94.9 mol%) (2.68 mPa·s, 353.15 K) at 0.1 MPa, respectively.

Table 1. CO_2 solubility of ionic liquid (IL)-based hybrid solvents.

IL-Based Hybrid Solvents	T/K	P/MPa	CO_2 Solubility (mol CO_2/kg Absorbent)	CO_2 Solubility (mol CO_2/mol IL)	Ref.
IL–H_2O					
[DMAPAH][Formate] (0.5:1, 1.0:1, 2.0:1, 2.5:1) + H_2O (20, 33, 42, 49, 59, 75 wt %)	298.15	0.1	0.41–0.14, 2.73–0.77, 4.23–1.41, 4.61–1.41		[30]
[P_{4442}][Suc] + H_2O (3.3, 8.8, 17.6 wt %)	293	0.1		1.9, 1.24, 0.81	[31]
[TMGH][Im] + H_2O (1, 2, 3, 5, 7, 10, 15, 20, 25 wt %)	313.15	0.1	3.39–4.23		[12]
[P_{4444}][HCOO] + H_2O (29, 32, 38, 50, 66, 70, 73, 79, 86, 91 mol%)	248.75–333.15	0.1		0.01–1	[32]
[BMMIM][Im] + H_2O (67, 91, 99, 99.9 mol%)	/	2	1.29, 2.30, 0.83, 0.55	0, 0, 0.36, 8.15	[33]
[BMMIM][Im] + H_2O (99 mol%)	/	1		0.45	[33]
[BMMIM][Ac] + H_2O (67, 99.9 mol%)	/	2		0, 6.95	[33]
[P_{4443}][Gly] + H_2O (59.9, 80.1, 90, 95 wt %)	278.09–348.10	0.103–7.53	0.091–2.44	0.23–12.13	[34]
[HMIM][Tf$_2$N] + H_2O (0.1, 1, 2, 5, 10 wt %)	298–333	2		0.30–0.13	[35]
[HMIM][FAP] + H_2O (0.1, 1, 2, 5, 10 wt %)	298–333	2		0.48–0.14	[35]
[BMIM][Ac] + H_2O (0.1, 1, 2, 5, 10 wt %)	298–323	2		0.46–0.32	[35]
[Chol][Gly] + H_2O (70 wt %)	308.15–338.15	0.0046–0.68	0.36–1.24	0.21–0.74	[17]
IL–organic/organic aqueous solution					
[TETAH][Lys] + ethanol + H_2O (H_2O:ethanol = 8:2, 6:4, 5:5, 4:6, 3:7, 2:8 $v:v$)	303	0.1		2.45–1.53	[36]
[AMIM][Tf$_2$N] + methanol (20, 50, 80 wt %)	313.2–353.2	0.98–6.19		2.15–3.89	[39]
[Chol][Gly] (30 wt %) + PEG200 (10 wt%) + H_2O (60 wt %)	308.15–338.15	0.0054–0.68	0.37–1.22	0.22–0.72	[17]
[Chol][Gly] (30 wt %) + PEG200 (20 wt%) + H_2O (50 wt %)	308.15–338.15	0.0039–0.68	0.41–1.21	0.24–0.72	[17]
[Chol][Gly] (30 wt %) + PEG200 (30 wt%) + H_2O (40 wt %)	308.15–338.15	0.0065–0.68	0.41–1.23	0.24–0.73	[17]
[Chol][Pro] + PEG200 (50, 67, 75 wt %)	308.15–353.15	0.0041–0.11		0.099–0.61	[40]
[P_{4444}][Gly] + PEG400 (70 wt %)	333.15–413.15	0.088–1.7	0.19–1.58	0.19–1.23	[41]
[P_{4444}][Ala] + PEG400 (70 wt %)	333.15–413.15	0.093–1.7	0.11–1.57	0.11–1.26	[41]
[P_{4444}][Pro] + PEG400 (70 wt %)	333.15–413.15	0.096–1.71	0.15–1.61	0.17–1.41	[41]
[DETAH][Br] + PEG200 (80 wt %)	293.15	0.1		1.18	[42]
[DETAH][Br] + PEG300 (80 wt %)	293.15	0.1		0.87	[42]
[DETAH][Br] + PEG400 (80 wt %)	293.15	0.1		0.32	[42]
[DETAH][BF$_4$] + PEG200 (80 wt %)	293.15	0.1		0.65	[42]
[DETAH][Br] + PEG200 + H_2O (1.3, 4.7 wt %)	293.15	0.02–0.5		1.05, 1.18	[42]
[N$_{1114}$][Tf$_2$N] + PEO1000 (10.44, 28.27, 50.22, 75.31 mol%)	323, 343			0.0057–1.16	[43]
[BMIM][BF$_4$] + TEG (20, 50, 80 mol%)	273.15–353.15	0.42–3.55		0.051–1.72	[44]

Table 1. Cont.

IL-Based Hybrid Solvents	T/K	P/MPa	CO_2 Solubility (mol CO_2/kg Absorbent)	CO_2 Solubility (mol CO_2/mol IL)	Ref.
[BMIM][BF$_4$] (56 mol%) + TEG (14 mol%) + H$_2$O (30 mol%)	293.15–333.15	0.38–4.17		0.051–0.96	[44]
[BMIM][BF$_4$] (35 mol%) + TEG (35 mol%) + H$_2$O (30 mol%)	293.15–333.15	0.57–4.37		0.079–1.4	[44]
[BMIM][BF$_4$] (14 mol%) + TEG (56 mol%) + H$_2$O (30 mol%)	293.15–333.15	0.64–4.46		0.15–1.84	[44]
[P$_{66614}$][3-Triz] + TG (30 mol%)	313.15–353.6	0.037–2.75		0.06–1.55	[18]
[P$_{66614}$][4-Triz] + TG (30 mol%)	313.15 353.6	0.07 3.03		0.075–2.23	[18]
[TEPAH][2-MI] + NPA + EG	303.15	0.1		1.72	[45]
IL-amine/amine aqueous solution					
[BMPyrr][DCA] (5 wt %) + DEA (35 wt%) + H$_2$O (60 wt %)	333.15	0.5–0.7		0.19 0.71	[52]
[BMPyrr][DCA] (10 wt %) + DEA (30wt%) + H$_2$O (60 wt %)	333.15	0.5–0.7		0.58–0.81	[52]
[DMAPAH][Formate] (1.0:1, 2.5:1) + MEA	298.15	0.1	3.82, 4.52		[30]
[DMEDAH][Formate] (1.0:1, 2.5:1) + MEA	298.15	0.1	3.52, 4.07		[30]
[DMAPAH][Formate] (0.5:1, 1.0:1, 2.0:1, 2.5:1) + MEA (30 wt %)	298.15	0.1	2.85, 4.57, 6.24, 5.89		[30]
[DMAPAH][Octanoate] (0.5:1, 1.0:1, 2.0:1, 2.5:1) + MEA (30 wt %)	298.15	0.1	2.29, 2.69, 3.30, 3.65		[30]
[DMAPAH][Ac] + MDEA (20, 33, 43, 50, 60, 67, 80 mol%)	308.15	0.0–3.0		0–3.13	[50]
[DMAPAH][Ac] + MDEA (50 mol%)	298.15–328.15	0.0–3.0	0–3.21		[50]
[N$_{1111}$][Lys] + DMEE (95, 90, 80, 60, 40 wt %)	303	0.1	0.28–1.69	1.22–0.61	[51]
[N$_{1111}$][Lys] + DMEE (80 wt %)	313, 323	0.1		0.76, 0.72	[51]

The CO_2 solubility unit with mole scale for reference [45] is mol CO_2/mol absorbent, while in [52] is mol CO_2/mol amine.

Table 2. Henry's constant of IL-based hybrid solvents.

IL-Based Hybrid Solvents	T (K)	Henry's Constant (MPa)	Ref.
IL–H$_2$O			
[P$_{4443}$][Gly] + H$_2$O (59.9, 80.1, 90%, 95 wt %)	278.14–348.05	2.8–5.05, 1.53–3.3, 0.87–2.81, 0.35–1.03	[34]
[Cho][Gly] + H$_2$O (70 wt %)	308.15–338.15	40.56–58	[17]
IL–organic			
[Cho][Gly] (30 wt %) + PEG200 (10 wt %) + H$_2$O (60 wt %)	308.15–338.15	36.6–52.2	[17]
[Cho][Gly] (30 wt %) + PEG200 (20 wt %) + H$_2$O (50 wt %)	308.15–338.15	33–47.03	[17]
[Cho][Gly] (30 wt %) + PEG200 (30 wt %) + H$_2$O (40 wt %)	308.15–338.15	31.49–46.79	[17]
[BMIM][BF$_4$] + TEG (20, 50, 80 mol%)	273.15–353.15	4.84–20.02, 5.65–22, 6.48–27.63	[44]

Table 3. Viscosity of IL-based hybrid solvents.

IL-Based Hybrid Solvents	T/K	P/MPa	Viscosity (mPa·s)	Ref.
IL-H$_2$O				
[DEA][Bu] + H$_2$O (98.78, 95.49, 90.01, 85.7, 79.49, 69.36, 51.21, 33.16, 2.6 mol%)	283.15–343.15	0.1	2.24–0.59, 6.91–1.10, 19.26–2.25, 32.03–3.24, 56.72–4.81, 106.48–7.09, 158.58–9.28, 158.29–9.64, 130.18–8.70	[14]
[P$_{4444}$][HCOO] + H$_2$O (25, 50, 60, 74, 80, 91 mol%)	298.15	0.1	356, 146, 97, 48, 35.3, 14.4	[32]
[HMIM][Tf$_2$N] + H$_2$O (0.1, 1, 2, 5, 10 wt %)	298–333	2	69.18–18.6, 66.54–18.2, 63.71–17.54, 55.95–15.7, 45.05–13.01	[35]
[HMIM][FAP] + H$_2$O (0.1, 1, 2, 5, 10 wt %)	298–333	2	88.09–20.71, 84.54–20.70, 80.76–20.66, 70.4–20.6, 56.01–20.47	[35]
[BMIM][Ac] + H$_2$O (0.1, 1, 2, 5, 10 wt %)	298–333	2	389–45.30, 225.86–30.04, 135.5–20.45, 44.08–8.78, 14.34–3.77	[35]
IL–organic/organic aqueous solution				
[TETAH][Lys] + ethanol + H$_2$O (H$_2$O:ethanol = 8:2, 6:4, 5:5, 4:6, 3:7, 2:8 v:v)	303	0.1	2.57, 3.00, 3.50, 3.81, 3.74, 3.51	[36]
[Chol][Gly] + PEG200 (70 wt %)	308.15–338.15	0.1	101.3–28.6	[17]
[Chol][Gly]/H$_2$O + PEG200 (30 wt %)	308.15, 338.15	0.1	7.96, 3.43	[17]
[P$_{4444}$][Gly] + PEG400 (70 wt %)	298.15–393.15	0.1	180.47–8.96	[41]
[P$_{4444}$][Ala] + PEG400 (70 wt %)	298.15–393.15	0.1	216.64–9.13	[41]
[P$_{4444}$][Pro] + PEG400 (70 wt %)	298.15–393.15	0.1	481–14.1	[41]
[DETAH][Br] + PEG200 (80 mol%)	293.15	0.1	71.7	[42]
[N$_{1114}$][Tf$_2$N] + PEO222 (26.23, 50.07, 75.06 mol%)	293.15–353.15	0.1	135.22–2.57	[43]
[N$_{1114}$][Tf$_2$N] + PEO500 (20.90, 44.22, 70.40 mol%)	293.15–353.15	0.1	135.22–7.51	[43]
[N$_{1114}$][Tf$_2$N] + PEO1000 (25.04, 50.22, 75.31 mol%)	318.15–353.15	0.1	51.72–17.22	[43]
[P$_{6614}$][4-NO$_2$imid] + TG (18.9, 40.4 mol%)	278.15–323.15	0.1	1503–57	[18]
[P$_{6614}$][4,5-CNimid] + TG (20.1, 40.1 mol%)	278.15–323.15	0.1	1051–63	[18]
[P$_{6614}$][Tf$_2$N] + TG (19.1, 36.6, 55.9, 66.4 mol%)	278.15–323.15	0.1	589–51	[18]
[P$_{6614}$][2-CH$_3$,5-NO$_2$imid] + TG (15.1, 39.9 mol%)	278.15–323.15	0.1	2535–57	[18]
[P$_{6614}$][DCA] + TG (9.9, 20.3, 29.9, 50.1, 65.2 mol%)	278.15–323.15	0.1	1130–54	[18]
[HMIM][Tf$_2$N] + TG (9.9, 19, 27.6, 41.3, 54.4 mol%)	278.15–293.15	0.1	171–50	[18]
[P$_{6614}$][BrBnIm] + TG (10.2, 13.8, 21.8, 34.6, 47.8, 63.8 mol%)	278.15–323.15	0.1	4400–51	[18]
[P$_{6614}$][Ac] + TG (10.3, 20.7, 29.9, 40, 49.9 mol%)	278.15–323.15	0.1	1130–57	[18]
[P$_{6614}$][Tf$_2$N] + TG (10.2, 20.5, 30.6, 39.9, 49.9, 59.9 mol%)	278.15–323.15	0.1	370–48	[18]
[HMMIM][Tf$_2$N] + TG (4.6, 12.6, 19.6, 24.9, 30.6, 39, 48.4, 58.2 mol%)	278.15–323.15	0.1	3290–53	[18]
[P$_{6614}$][3-Triz] + TG (8.3, 12.2, 20.7, 31, 39.8, 50.1, 70 mol%)	278.15–323.15	0.1	1180–51	[18]
[P$_{44412}$][3-Triz] + TG (6.2, 12.3, 15.5, 21.4, 30.5, 39.5, 49.5, 59.6 mol%)	278.15–323.15	0.1	1980–57	[18]

Table 3. Cont.

IL-Based Hybrid Solvents	T/K	P/MPa	Viscosity (mPa·s)	Ref.
$[P_{2228}][4-NO_2pyra]$ + TG (5.1, 10.2, 19.9, 30, 40.4, 50.3 mol%)	278.15–323.15	0.1	1010–55	[18]
$[P_{2228}][4-NO_2imid]$ + TG (10, 20, 30.1, 40.1, 50.1 mol%)	278.15–323.15	0.1	700–55	[18]
$[P_{2228}][2-CH_3,5-NO_2imid]$ + TG (3.6, 6.7, 11.5, 23.3, 30, 39.9, 50, 59.8 mol%)	278.15–323.15	0.1	2730–51	[18]
[mm(butene)im][4-NO_2pyra] + TG (4.8, 10, 19.9, 29.9, 39.9, 50, 60 mol%)	278.15–323.15	0.1	4300–51	[18]
$[P_{2224}][2-CH_3,5-NO_2imid]$ + TG (4.8, 10.2, 20.1, 29.9, 40.1, 50.1 mol%)	278.15–323.15	0.1	1540–57	[18]
[pmmim][4-NO_2pyra] + TG (5, 10, 20.1, 30.2, 39.9, 49.9, 60 mol%)	278.15 323.15	0.1	5120 59	[18]
[TEPAH][2-MI] + NPA + EG	303.15	0.1	3.66	[45]
IL–amine				
$[N_{1111}][Lys]$ + DMEE (95, 90, 80, 60, 40 wt %)	303–333	0.1	12.09–4.88, 20.70–6.98, 30.00–9.30, 80.46–20, 101.86–25.35	[51]
$[BMIM][BF_4]$ + DETA (94.9, 80.14, 70.02, 60.05, 50.08, 40.95, 30.26, 19.88, 10.52, 5.04 mol%)	298.15–333.15	0.1	6.71–2.68, 12.1–4.89, 17.35–6.69, 23.66–8.72, 31.54–11.19, 40.07–13.88, 52.59–17.57, 67.10–21.30, 82.58–24.18, 91.94–24.54	[53]

3. DESs-Based Hybrid Solvents

The CO_2 solubility data for 33 kinds of DESs-based hybrid solvents, together with viscosities for six types of DESs-based hybrid solvents since 2016, and Henry's constants for 21 kinds of DES-based hybrid solvents since 2013 have been reported, as summarized in Tables 4 and 5. The full names of the studied components of DESs are given in Table S1.

3.1. CO_2 Solubility

3.1.1. DES–H_2O

The CO_2 solubility of DESs using water as a hybrid solvent was investigated by Sarmad et al. at 298.15 K and pressure up to 2 MPa (Table 4) [9]. From this study, the CO_2 solubility of DES-based hybrid solvents can be affected by four factors. (1) The first is pressure: the CO_2 solubility increased with the increasing pressure. For instance, the CO_2 solubility of [TEMA][Cl]-GLY-H_2O 1:2:0.11 increased from 0.025 to 0.66 mol/kg at a pressure range of 0.14 to 1.74 MPa. (2) The second factor is the mole ratio of water: the CO_2 solubility of [TEMA][Cl]-GLY-H_2O 1:2:0.05 first decreased at pressure range of 0.23–0.85 MPa, and then, it increased from 1.25 to 1.98 MPa compared with [TEMA][Cl]-GLY 1:2. Upon increasing the mole ratio of water, the CO_2 solubility of [TEMA][Cl]-GLY-H_2O 1:2:0.11 significantly enhanced with respect to [TEMA][Cl]-GLY 1:2 and [TEMA][Cl]-GLY-H_2O 1:2:0.05. This result agrees with the DES-based hybrid solvent of [BTMA][Cl]-GLY-H_2O 1:2:0.011 and 1:2:0.05. (3) The third factor is the type of hydrogen bond acceptor (HBA)—for example, the CO_2 solubility of [TEMA][Cl]-GLY-H_2O 1:2:0.05 (1.98 MPa, 0.66 mol/kg) > [BTMA][Cl]-GLY-H_2O 1:2:0.05 (2.02 MPa, 0.29 mol/kg).

Harifi-Mood et al. investigated the Henry's constants for [Ch][Cl]-EG aqueous solution at temperatures of 303.15–323.15 K [54]. As shown in Table 5, the results indicate that the Henry's constant of CO_2 increases with increasing water amount in the absorbent, corresponding to a decrease of CO_2 solubility. This result agrees with the measured Henry's constants of [Ch][Cl]-EG, [Ch][Cl]-GLY, and [Ch][Cl]-MA aqueous solution from 303.15 to 313.15 K by Lin et al. [55].

3.1.2. DES–Organic

The CO_2 solubilities in DESs–organic hybrid solvents are given in Table 4. The result indicates that the addition of 0.03 mol of acetic acid in [MTPP][Br]-LEV 1:3 significantly enhanced the CO_2 solubility and decreased the viscosity compared with [MTPP][Br]-LEV 1:3 [9].

A superbase can participate in the reaction of DES and CO_2, thus increasing the CO_2 solubility. Bhawna et al. studied the CO_2 solubility by three hybrid superbases of TBD, DBN, and DBU with DESs of [Ch][Cl]-Urea 1:2 and [Ch][Cl]-EG 1:2, respectively [25]. The result indicates that all of these three superbases can enhance the CO_2 solubility, and among them, TBD has the highest capacity, followed by DBU and DBN. The further addition of glycerol in these hybrid solvents decreased the CO_2 solubility. For the effect of these three superbases on different male ratio of DESs, it is found that [Ch][Cl]-MEA 1:2 + DBN (5.11 mol/kg) > [Ch][Cl]-MEA + TBD 1:4 (3.91 mol/kg) > [Ch][Cl]-MEA 1:2 + DBU (3.54 mol/kg). Additionally, the same phenomenon was observed that the addition of glycerol in these [Ch][Cl]-MEA-based hybrid solvents can decrease the CO_2 solubility.

An imidazole (Im)-derived DESs of [BMIM][Cl]-Im was synthesized for CO_2 capture by hybrid with DBN [56]. These hybrid solvents show remarkable CO_2 capture capacity up to 1.00 mol/mol, following the order of DBN-[BMIM][Cl]-Im 1:1:2 > DBN-[BMIM][Cl]-Im 1:1:1 > DBN-[BMIM][Cl]-Im 1:2:1. The theoretical calculation indicates that DBN plays a key role in the absorption process by forming a strong hydrogen bond with the derived [$BMIM^+$-COO^-].

In conclusion, the obtained best DES–H_2O hybrid solvent is [TEMA][Cl]-GLY-H_2O 1:2:0.11 (0.66 mol/kg, 298 K, 1.74 MPa), while it is [Ch][Cl]-MEA 1:2 + DBN 1:1 (5.11 mol/kg, 298 K, 0.1 MPa) for DES–organic hybrid solvent. These values are lower than the best IL–H_2O and IL–organic hybrid solvents, respectively.

Table 4. CO_2 solubilities and viscosities of deep eutectic solvent (DES)-based hybrid solvents.

DES-Based Hybrid Solvents	Amount/Ratio	Temperature (K)	Pressure (MPa)	CO_2 Solubility (mol CO_2/kg DES)	CO_2 Solubility (mol CO_2/mol DES)	Viscosity (mPa·s)	Ref.
\multicolumn{8}{c}{DES–H_2O}							
[BHDE][Cl]-GLY-H_2O	1:3:0.11	298.15	0.23–2.02	0.037–0.21		32.76–17.11	[9]
[BTMA][Cl]-GLY-H_2O	1:2:0.05	298.15	0.21–2.02	0.044–0.29		70.76–26.81	[9]
[BTMA][Cl]-GLY-H_2O	1:2:0.11	298.15	0.26–2.03	0.016–0.33		22.19–14.41	[9]
[TEMA][Cl]-GLY-H_2O	1:2:0.05	298.15	0.23–1.98	0.009–0.66		90.98–23.56	[9]
[TEMA][Cl]-GLY-H_2O	1:2:0.11	298.15	0.14–1.74	0.025–0.66		48.63–19.31	[9]
[L-Arg]-GLY 1:6-H_2O	10, 20, 30, 40, 50, 60 wt %	303.15–353.15	0.1			434.2–26.3, 179.3–19.5, 45.2–7.2, 17.7–4.4, 8.0–3.6, 4.6–3.0	[57]
\multicolumn{8}{c}{DES–organic}							
[Ch][Cl]-GLY-AC	1:1:1	298.15	0.26–2.01	0.052–0.43		138.51–28.81	[9]
[MTPP][Br]-LEV-AC	1:3:0.03	298.15	0.29–2.06	0.18–1.32		40.12–16.16	[9]
[Ch][Cl]-Urea 1:2 + TBD	1:10	333.15	0.1	0.42	0.68		[25]
[Ch][Cl]-Urea 1:2 + DBU	1:10	333.15	0.1	0.56	1.21		[25]
[Ch][Cl]-Urea 1:2 + DBN	1:10	333.15	0.1	0.76	1.11		[25]
[Ch][Cl]-Urea 1:2-GLY + TBD	5:40:10	333.15	0.1	0.4	0.66		[25]
[Ch][Cl]-Urea 1:2-GLY + DBU	5:40:10	333.15	0.1	0.45	0.81		[25]
[Ch][Cl]-Urea 1:2-GLY + DBN	5:40:10	333.15	0.1	0.27	0.40		[25]
[Ch][Cl]-EG 1:2 + TBD	1:10	333.15	0.1	0.7	1.16		[25]
[Ch][Cl]-EG 1:2 + DBN	1:10	333.15	0.1	0.77	1.06		[25]
[Ch][Cl]-EG 1:2 + TBD	1:10	333.15	0.1	0.63	0.95		[25]
[Ch][Cl]-EG 1:2-GLY + DBU	5:40:10	333.15	0.1	0.68	1.16		[25]
[Ch][Cl]-EG 1:2 + DBU	1:10	298.15	0.1	0.83	1.41		[25]
[Ch][Cl]-EG 1:2 + DBN	1:10	298.15	0.1	0.86	1.19		[25]
[Ch][Cl]-EG 1:2 + TBD	1:10	298.15	0.1	0.76	1.17		[25]
[Ch][Cl]-EG 1:2-GLY + DBU	5:40:10	298.15	0.1	0.64	1.12		[25]
[Ch][Cl]-EG 1:2-GLY + DBN	5:40:10	298.15	0.1	0.74	1.05		[25]
[Ch][Cl]-EG 1:2-GLY + TBD	5:40:10	298.15	0.1	0.73	1.17		[25]
[Ch][Cl]-MEA 1:2 + DBU	1:10	298.15	0.1	3.54	2.86		[25]
[Ch][Cl]-MEA 1:2 + DBN	1:10	298.15	0.1	5.11	6.70		[25]
[Ch][Cl]-MEA 1:4 + TBD	1:10	298.15	0.1	3.91	5.77		[25]
[Ch][Cl]-MEA 1:2-GLY + DBU	5:40:10	298.15	0.1	3.26	5.46		[25]

Table 4. Cont.

DES-Based Hybrid Solvents	Amount/ Ratio	Temperature (K)	Pressure (MPa)	CO_2 Solubility (mol CO_2/kg DES)	CO_2 Solubility (mol CO_2/mol DES)	Viscosity (mPa·s)	Ref.
[Ch][Cl]-MEA 1:2-GLY + DBN	5:40:10	298.15	0.1	1.67	2.28		[25]
[Ch][Cl]-MEA 1:2-GLY + TBD	5:40:10	298.15	0.1	3.63	5.56		[25]
[BMIM][Cl]-Im + DBN	1:1:1	298.15–328.15	0.1		0.81–1.02		[56]
[BMIM][Cl]-Im + DBN	1:1:2	298.15–328.15	0.1		0.88–0.97		[56]
[BMIM][Cl]-Im + DBN	1:2:1	298.15–328.15	0.1		0.91–1.07		[56]

The ratio of DES–organic for reference [9] is mole ratio, while in [25] is volume ratio.

Table 5. Henry's constant of DES–H_2O hybrid solvents.

DES–H_2O	T (K)	Henry's Constant	Ref.
[Ch][Cl]-EG + H_2O (10, 20, 30, 40, 50, 60, 70, 80, 90 mol%)	303.15–313.15	27–296	[54]
[Ch][Cl]-EG + H_2O (20, 40, 60, 80 wt %)	303.15–313.15	3965.5–2805.5	[55]
[Ch][Cl]-GLY + H_2O (20, 40, 60, 80 wt %)	303.15–313.15	3818.8–3185.2	[55]
[Ch][Cl]-MA + H_2O (20, 40, 60, 80 wt %)	303.15–313.15	4021.6–2946.2	[55]

The Henry's constant unit in reference of [54] is MPa, while it is kPa·m^3·kmol^{-1} for reference [55].

3.2. Viscosity

The DESs consisting of glycerol as the hydrogen bond donor (HBD) exhibited high viscosity. Meanwhile, their viscosities increased considerably with an increase in the amount of dissolved CO_2. As shown in Table 4, using water as a hybrid solvent in glycerol-based DESs can significantly decrease the viscosity of the DES [9]. For example, the viscosity of [BTMA][Cl]-GLY 1:2 decreased from 1017.67 to 70.76 mPa·s when adding a 0.05 molar ratio of water in [BTMA][Cl]-GLY 1:2 (i.e., [BTMA][Cl]-GLY-H_2O 1:2:0.05), but limiting the contribution of H_2O to CO_2 solubility. Meanwhile, increasing the water content of 0.11 mol in [BTMA][Cl]-GLY 1:2 results in a considerably reduced viscosity, which agrees with the results in the [TEMA][Cl]-GLY-H_2O system. Additionally, the addition of 0.11 mol of water to the DES of [BHDE][Cl]-GLY 1:3 decreased the viscosity from 32.76 to 17.11 mPa·s at 298.15 K and 0.23–2.02 MPa, and it increased the CO_2 solubility from 0.037 to 0.21 mol/kg. The viscosity of the [L-Arg]-GLY 1:6 as a function of water content from 0 to 60 wt % was measured, which indicates that viscosity of the DES decreased sharply with the increase of water contents, giving an option to lower the viscosity [57].

In a word, adding water and organic solvents in DES can significantly decrease the viscosity.

4. Comparison of CO_2 Solubility and Viscosity

The obtained best candidates of IL–H_2O, IL–organic, IL–amine, DES–H_2O, and DES–organic hybrid solvents were compared with each other and their pure ILs and DESs (Figure 4). As shown in Figure 4, for either the IL-based or DES-based hybrid solvents, their CO_2 solubilies are higher than their pure IL/DES under the same condition. For example, the CO_2 solubility of [DMAPAH][Formate] (2.5:1) + H_2O (20 wt %) is 4.61 mol/kg at 298 K and 0.1 MPa, while it is 2.32 mol/kg for [DMAPAH][Formate] (2.5:1). This result indicates that IL/DES-based hybrid solvents are remarkable ones for CO_2 capture. Additionally, the IL-based hybrid solvent shows better CO_2 capture performance compared with the DES-based hybrid solvent, as shown in Figure 4. Figure 5 gives the comparison of viscosities for these IL/DES-based hybrid solvents and pure IL and DES at 333.15 K and 0.1 MPa. As shown in Figure 5, the addition of hybrid solvents can significantly decrease the viscosity compared to pure ILs and DESs, which are beneficial to accelerate mass transfer during capturing CO_2.

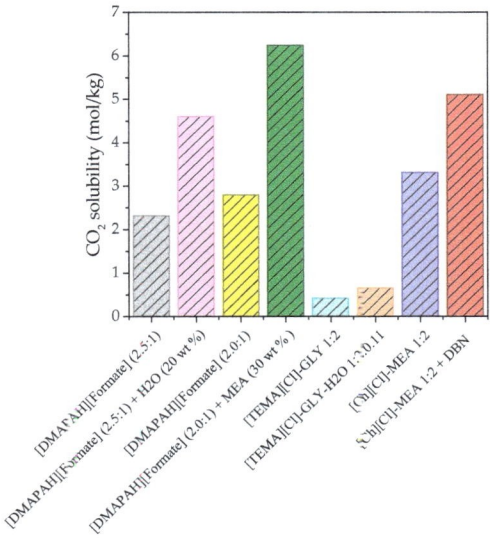

Figure 4. Comparison of CO_2 solubility for IL and IL-based hybrid solvent, as well as DES and DES-based hybric solvent.

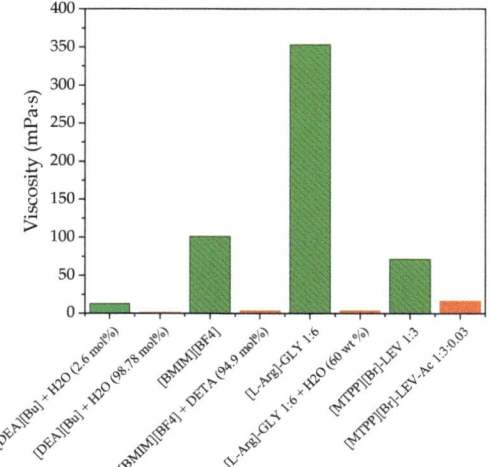

Figure 5. Comparison of viscosity for IL and IL-based hybrid solvent, as well as DES and DES-based hybrid solvent at 333.15 K and 0.1 MPa.

Viscosity is the key factor for impeding the mass transfer of gas in absorbent [58]. For example, Gómez-Coma et al. investigated the viscosity and mass transfer performance of [EMIM][Ac] + H_2O for CO_2 capture [59], finding that the viscosity of [EMIM][Ac] + H_2O decreased from 150 to 20 mPa·s with the increase of water content from 0–40 wt %. For the mass transfer coefficient, it is first increased from 1.7×10^{-5} to 9.34×10^{-5} m·s^{-1} with the increasing of water from 0 to 30 wt % in [EMIM][Ac] and then decreased to 6.81×10^{-5} m·s^{-1} when water content up to 40 wt %. Huang et al. evidenced that the low viscosity of IL–MEA aqueous solution corresponds to a high mass transfer performance [60], i.e., [EMIM][Br] (20 wt %) + MEA (5 wt %) + H_2O (75 wt %) (11.57×10^6 mol·m^{-3}·s^{-1}·Pa^{-1}, 1.23 mPa·s) > [BMIM][Br] (20 wt %) + MEA (5 wt %) + H_2O (75 wt %) (11.04×10^6 mol·m^{-3}·s^{-1}·Pa^{-1}, 1.3 mPa·s) > [EMIM][Br] (30 wt %) + MEA (5 wt %) + H_2O (65 wt %) (9.86×10^6 mol·m^{-3}·s^{-1}·Pa^{-1}, 1.42 mPa·s) > [BMIM][Br] (30 wt %) + MEA (5 wt %) + H_2O (65 wt %) (9.67×10^6 mol·m^{-3}·s^{-1}·Pa^{-1}, 1.6 mPa·s). A similar phenomenon can be found in DESs hybrid solvent. Ma et al. indicated that a small amount of water in [BTMA][Cl]-GLY 1:2 not only decreases the viscosity but also improves the CO_2 solubility due to the increase of the mass transfer [61], while excess water in [BTMA][Cl]-GLY 1:2 results in a decrease of CO_2 solubility, which is in agreement with Li et al. [12].

5. Conclusions

This review summarizes the research work on developing ILs/DESs-based hybrid solvents (i.e., IL–H_2O, IL–organic/organic aqueous solution, IL–amine, DES–H_2O, and DES–organic) for CO_2 capture, including CO_2 solubility, Henry's constant, and viscosity. The results illustrate that the addition of hybrid solvents to ILs and DESs can decrease the viscosity and enhance the CO_2 solubility. IL–amine based hybrid solvents are super to IL–H_2O and IL–organic/organic aqueous solution, and some of the IL-based hybrid solvents show better performance than that of DES-based hybrid solvents. Additionally, some of the IL/DES hybrid solvents have higher CO_2 solubility compared to their pure IL/DES, indicating that the addition of hybrid solvent to IL/DES is possible to develop greener and more efficient absorbents for CO_2 capture. To develop the efficient IL/DES hybrid solvents for CO_2 capture, the following aspects are suggested for consideration to decrease the viscosity and increase the CO_2 solubility: (1) hybrid of functional ILs/DESs that have high CO_2 solubilities with a certain amount of water; (2) the addition of organic solvent which has a small molecular weight to the ILs/DESs; and (3) applying amine solvent which has good CO_2 capture capacity to ILs and DESs.

Supplementary Materials: The following is available online at http://www.mdpi.com/2073-4352/10/11/978/s1, Table S1: Full names and abbreviations of ILs, components of DESs and hybrid solvents.

Author Contributions: Writing—original draft preparation, Y.L.; investigation, Z.D.; writing—review and editing, F.D.; conceptualization and supervision, X.J. All authors have read and agreed to the published version of the manuscript.

Funding: This work is financially supported by Carl Tryggers Stiftelse foundation (No. 18:175). X.J. thanks the financial support from the Swedish Energy Agency (No. P47500-1) and K. C. Wang Education Foundation (No. GJTD-2018-04). F.D. thanks the financial support from the National Nature Science Foundation of China (21808223).

Conflicts of Interest: The author declares no conflict of interest.

References

1. Zhou, W.; Wang, J.; Chen, P.; Ji, C.; Kang, Q.; Lu, B.; Li, K.; Liu, J.; Ruan, R. Bio-mitigation of carbon dioxide using microalgal systems: Advances and perspectives. *Renew. Sustain. Energy Rev.* **2017**, *76*, 1163–1175. [CrossRef]
2. MacDowell, N.; Florin, N.; Buchard, A.; Hallett, J.; Galindo, A.; Jackson, G.; Adjiman, C.S.; Williams, C.K.; Shah, N.; Fennell, P. An overview of CO_2 capture technologies. *Energy Environ. Sci.* **2010**, *3*, 1645. [CrossRef]
3. Oko, E.; Zacchello, B.; Wang, M.H.; Fethi, A. Process analysis and economic evaluation of mixed aqueous ionic liquid and monoethanolamine (MEA) solvent for CO_2 capture from a coke oven plant. *Greenh. Gases* **2018**, *8*, 686–700. [CrossRef]
4. Kothandaraman, A.; Nord, L.; Bolland, O.; Herzog, H.J.; McRae, G.J. Comparison of solvents for post-combustion capture of CO_2 by chemical absorption. *Energy Procedia* **2009**, *1*, 1373–1380. [CrossRef]
5. Lawal, A.; Wang, M.; Stephenson, P.; Obi, O. Demonstrating full-scale post-combustion CO_2 capture for coal-fired power plants through dynamic modelling and simulation. *Fuel* **2012**, *101*, 115–128. [CrossRef]
6. Ramdin, M.; de Loos, T.W.; Vlugt, T.J.H. State-of-the-art of CO_2 capture with ionic liquids. *Ind. Eng. Chem. Res.* **2012**, *51*, 8149–8177. [CrossRef]
7. Zeng, S.J.; Zhang, X.P.; Bai, L.; Zhang, X.C.; Wang, H.; Wang, J.J.; Bao, D.; Li, M.D.; Liu, X.Y.; Zhang, S.J. Ionic-liquid-based CO_2 capture systems: Structure, interaction and process. *Chem. Rev.* **2017**, *117*, 9625–9673. [CrossRef] [PubMed]
8. Huang, Y.J.; Cui, G.K.; Wang, H.Y.; Li, Z.Y.; Wang, J.J. Tuning ionic liquids with imide-based anions for highly efficient CO_2 capture through enhanced cooperations. *J. CO2 Util.* **2018**, *28*, 299–305. [CrossRef]
9. Sarmad, S.; Xie, Y.J.; Mikkola, J.-P.; Ji, X.Y. Screening of deep eutectic solvents (DESs) as green CO_2 sorbents: From solubility to viscosity. *New J. Chem.* **2017**, *41*, 290–301. [CrossRef]
10. Liu, H.J.; Pan, Y.; Yao, H.; Zhang, Y. Enhancement of carbon dioxide mass transfer rate by (ionic liquid)-in-water emulsion. *Adv. Mater. Res.* **2014**, *881–883*, 113–117. [CrossRef]
11. Chu, C.Y.; Zhang, F.B.; Zhu, C.Y.; Fu, T.T.; Ma, Y.G. Mass transfer characteristics of CO_2 absorption into 1-butyl-3-methylimidazolium tetrafluoroborate aqueous solution in microchannel. *Int. J. Heat Mass Trans.* **2019**, *128*, 1064–1071. [CrossRef]
12. Li, F.F.; Bai, Y.G.; Zeng, S.J.; Liang, X.D.; Wang, H.; Huo, F.; Zhang, X.P. Protic ionic liquids with low viscosity for efficient and reversible capture of carbon dioxide. *Int. J. Greenh. Gas Control* **2019**, *90*, 102801. [CrossRef]
13. Zhang, X.; Bao, D.; Huang, Y.; Dong, H.F.; Zhang, X.P.; Zhang, S.J. Gas-liquid mass-transfer properties in CO_2 absorption system with ionic liquids. *AIChE J.* **2014**, *60*, 2929–2939. [CrossRef]
14. Alcantara, M.L.; Santos, J.P.; Loreno, M.; Ferreira, P.I.S.; Paredes, M.L.L.; Cardozo-Filho, L.; Silva, A.K. Lião, L.M.; Pires, C.A.M.; Mattedi, S. Low viscosity protic ionic liquid for CO_2/CH_4 separation: Thermophysical and high-pressure phase equilibria for diethylammonium butanoate. *Fluid Phase Equilibr.* **2018**, *459*, 30–43. [CrossRef]
15. Li, J.; You, C.J.; Chen, L.F.; Ye, Y.M.; Qi, Z.W.; Sundmacher, K. Dynamics of CO_2 Absorption and Desorption Processes in Alkanolamine with Cosolvent Polyethylene Glycol. *Ind. Eng. Chem. Res.* **2012**, *51*, 12081–12088. [CrossRef]
16. Yang, Z.-Z.; He, L.-N.; Zhao, Y.-N.; Li, B.; Yu, B. CO_2 capture and activation by superbase/polyethylene glycol and its subsequent conversion. *Energy Environ. Sci.* **2011**, *4*, 3971. [CrossRef]

17. Liu, S.D.; Li, H.; Chen, Y.F.; Yang, Z.H.; Wang, H.L.; Ji, X.Y.; Lu, X.H. Improved CO_2 separation performance of aqueous choline-glycine solution by partially replacing water with polyethylene glycol. *Fluid Phase Equilibr.* **2019**, *495*, 12–20. [CrossRef]
18. Fillion, J.J.; Bennett, J.E.; Brennecke, J.F. The viscosity and density of ionic liquid + tetraglyme mixtures and the effect of tetraglyme on CO_2 Solubility. *J. Chem. Eng. Data* **2017**, *62*, 608–622. [CrossRef]
19. Sairi, N.A.; Ghani, N.A.; Aroua, M.K.; Yusoff, R.; Alias, Y. Low pressure solubilities of CO_2 in guanidinium trifluoromethanesulfonate-MDEA systems. *Fluid Phase Equilibr.* **2015**, *385*, 79–91. [CrossRef]
20. Sairi, N.A.; Yusoff, R.; Alias, Y.; Aroua, M.K. Solubilities of CO_2 in aqueous N-methyldiethanolamine and guanidinium trifluoromethanesulfonate ionic liquid systems at elevated pressures. *Fluid Phase Equilibr.* **2011**, *300*, 89–94. [CrossRef]
21. Yang, J.; Yu, X.H.; Yan, J.Y.; Tu, S.-T. CO_2 capture using amine solution mixed with ionic liquid. *Ind. Eng. Chem. Res.* **2014**, *53*, 2790–2799. [CrossRef]
22. Xu, F.; Gao, H.S.; Dong, H.F.; Wang, Z.L.; Zhang, X.P.; Ren, B.Z.; Zhang, S.J. Solubility of CO_2 in aqueous mixtures of monoethanolamine and dicyanamide-based ionic liquids. *Fluid Phase Equilibr.* **2014**, *365*, 80–87. [CrossRef]
23. Gao, Y.; Zhang, F.; Huang, K.; Ma, J.-W.; Wu, Y.-T.; Zhang, Z.-B. Absorption of CO_2 in amino acid ionic liquid (AAIL) activated MDEA solutions. *Int. J. Greenh. Gas Control* **2013**, *19*, 379–386. [CrossRef]
24. Baj, S.; Siewniak, A.; Chrobok, A.; Krawczyk, T.; Sobolewski, A. Monoethanolamine and ionic liquid aqueous solutions as effective systems for CO_2 capture. *J. Chem. Technol. Biotechnol.* **2013**, *88*, 1220–1227. [CrossRef]
25. Bhawna; Pandey, A.; Pandey, S. Superbase-added choline chloride-based deep eutectic solvents for CO_2 capture and sequestration. *ChemistrySelect* **2017**, *2*, 11422–11430. [CrossRef]
26. Huang, K.; Chen, F.-F.; Tao, D.-J.; Dai, S. Ionic liquid–formulated hybrid solvents for CO_2 capture. *Curr. Opin. Green Sustain.* **2017**, *5*, 67–73. [CrossRef]
27. Zhang, F.; Fang, C.-G.; Wu, Y.-T.; Wang, Y.-T.; Li, A.-M.; Zhang, Z.-B. Absorption of CO_2 in the aqueous solutions of functionalized ionic liquids and MDEA. *Chem. Eng. J.* **2010**, *160*, 691–697.
28. Babamohammadi, S.; Shamiri, A.; Aroua, M.K. A review of CO_2 capture by absorption in ionic liquid-based solvents. *Rev. Chem. Eng.* **2015**, *31*, 383–412. [CrossRef]
29. Lian, S.H.; Song, C.F.; Liu, Q.L.; Duan, E.H.; Ren, H.W.; Kitamura, Y. Recent advances in ionic liquids-based hybrid processes for CO_2 capture and utilization. *J. Environ. Sci.* **2021**, *99*, 281–295. [CrossRef]
30. Vijayaraghavan, R.; Oncsik, T.; Mitschke, B.; MacFarlane, D.R. Base-rich diamino protic ionic liquid mixtures for enhanced CO_2 capture. *Sep. Purif. Technol.* **2018**, *196*, 27–31. [CrossRef]
31. Huang, Y.J.; Cui, G.K.; Zhao, Y.L.; Wang, H.Y.; Li, Z.Y.; Dai, S.; Wang, J.J. Reply to the correspondence on "Preorganization and cooperation for highly efficient and reversible capture of low-concentration CO_2 by ionic liquids". *Angew. Chem.* **2019**, *58*, 386–389. [CrossRef]
32. Yasaka, Y.; Kimura, Y. Effect of temperature and water concentration on CO_2 absorption by tetrabutylphosphonium formate ionic liquid. *J. Chem. Eng. Data* **2016**, *61*, 837–845. [CrossRef]
33. Simon, N.M.; Zanatta, M.; dos Santos, F.P.; Corvo, M.C.; Cabrita, E.J.; Dupont, J. Carbon dioxide capture by aqueous ionic liquid solutions. *ChemSusChem* **2017**, *10*, 4927–4933. [CrossRef]
34. Chen, Y.; Guo, K.H.; Huangpu, L. Experiments and modeling of absorption of CO_2 by amino-cation and amino-anion dual functionalized ionic liquid with the addition of aqueous medium. *J. Chem. Eng. Data* **2017**, *62*, 3732–3743. [CrossRef]
35. Aghaie, M.; Rezaei, N.; Zendehboudi, S. Assessment of carbon dioxide solubility in ionic liquid/toluene/water systems by extended PR and PC-SAFT EOSs: Carbon capture implication. *J. Mol. Liq.* **2019**, *275*, 323–337. [CrossRef]
36. Huang, Q.S.; Jing, G.H.; Zhou, X.B.; Lv, B.H.; Zhou, Z.M. A novel biphasic solvent of amino-functionalized ionic liquid for CO_2 capture: High efficiency and regenerability. *J. CO_2 Util.* **2018**, *25*, 22–30. [CrossRef]
37. Qian, Y.H.; Jing, G.H.; Lv, B.H.; Zhou, Z.M. Exploring the general characteristics of amino-acid-functionalized ionic liquids through experimental and quantum chemical calculations. *Energy Fuels* **2017**, *31*, 4202–4210. [CrossRef]
38. Zhou, X.B.; Jing, G.H.; Liu, F.; Lv, B.H.; Zhou, Z.M. Mechanism and kinetics of CO_2 absorption into an aqueous solution of a triamino-functionalized ionic liquid. *Energy Fuels* **2017**, *31*, 1793–1802. [CrossRef]
39. Taheri, M.; Dai, C.N.; Lei, Z.G. CO_2 capture by methanol, ionic liquid, and their binary mixtures: Experiments, modeling, and process simulation. *AIChE J.* **2018**, *64*, 2168–2180. [CrossRef]

40. Li, X.Y.; Hou, M.Q.; Zhang, Z.F.; Han, B.X.; Yang, G.Y.; Wang, X.L.; Zou, L.Z. Absorption of CO_2 by ionic liquid/polyethylene glycol mixture and the thermodynamic parameters. *Green Chem.* **2008**, *10*, 879. [CrossRef]
41. Li, J.; Dai, Z.D.; Usman, M.; Qi, Z.W.; Deng, L.Y. CO_2/H_2 separation by amino-acid ionic liquids with polyethylene glycol as co-solvent. *Int. J. Greenh. Gas Con.* **2016**, *45*, 207–215. [CrossRef]
42. Chen, Y.; Hu, H. Carbon dioxide capture by diethylenetriamine hydrobromide in nonaqueous systems and phase-change formation. *Energy Fuels* **2017**, *31*, 5363–5375. [CrossRef]
43. Lepre, L.F.; Pison, L.; Siqueira, L.J.A.; Ando, R.A.; Costa Gomes, M.F. Improvement of carbon dioxide absorption by mixing poly(ethylene glycol) dimethyl ether with ammonium-based ionic liquids. *Sep. Purif. Technol.* **2018**, *196*, 10–19. [CrossRef]
44. Jiang, Y.F.; Taheri, M.; Yu, G.Q.; Zhu, J.Q.; Lei, Z.G. Experiments, modeling, and simulation of CO_2 dehydration by ionic liquid, triethylene glycol, and their binary mixtures. *Ind. Eng. Chem. Res.* **2019**, *58*, 15588–15597. [CrossRef]
45. Liu, F.; Shen, Y.; Shen, L.; Sun, C.; Chen, L.; Wang, Q.L.; Li, S.J.; Li, W. Novel amino-functionalized ionic liquid/organic solvent with low viscosity for CO_2 capture. *Environ. Sci. Technol.* **2020**, *54*, 3520–3529. [CrossRef]
46. Huang, Q.; Li, Y.; Jin, X.B.; Zhao, D.; Chen, G.Z. Chloride ion enhanced thermal stability of carbon dioxide captured by monoethanolamine in hydroxyl imidazolium based ionic liquids. *Energy Environ. Sci.* **2011**, *4*, 2125. [CrossRef]
47. Fu, D.; Hao, H.M.; Liu, F. Experiment and model for the viscosity of carbonated 2-amino-2-methyl-1-propanol-monoethanolamine and 2-amino-2-methyl-1-propanol-diethanolamine aqueous solution. *J. Mol. Liq.* **2013**, *188*, 37–41. [CrossRef]
48. Goodrich, B.F.; de la Fuente, J.C.; Gurkan, B.E.; Zadigian, D.J.; Price, E.A.; Huang, Y.; Brennecke, J.F. Experimental measurements of amine-functionalized anion-tethered ionic liquids with carbon dioxide. *Ind. Eng. Chem. Res.* **2011**, *50*, 111–118. [CrossRef]
49. Liu, F.; Jing, G.H.; Zhou, X.B.; Lv, B.H.; Zhou, Z.M. Performance and mechanisms of triethylene tetramine (TETA) and 2-Amino-2-methyl-1-propanol (AMP) in aqueous and nonaqueous solutions for CO_2 capture. *ACS Sustain. Chem. Eng.* **2017**, *6*, 1352–1361. [CrossRef]
50. Zheng, W.-T.; Huang, K.; Wu, Y.-T.; Hu, X.-B. Protic ionic liquid as excellent shuttle of MDEA for fast capture of CO_2. *AIChE J.* **2018**, *64*, 209–219. [CrossRef]
51. Meng, Y.N.; Wang, X.D.; Zhang, F.; Zhang, Z.B.; Wu, Y.T. IL-DMEE Nonwater system for CO_2 capture: Absorption performance and mechanism investigations. *Energy Fuels* **2018**, *32*, 8587–8593. [CrossRef]
52. Salleh, R.M.; Jamaludin, S.N. Thermodynamic equilibrium solubility of diethanolamine–N-butyl-1-methylpyrrolidinium dicyanamide [DEABMPYRR DCA] mixtures for carbon dioxide capture. *IOP Conf. Ser. Mater. Sci. Eng* **2018**, *358*, 012010.53. [CrossRef]
53. Ahmad, W.; Al-Ami, A.; Vakili-Nezhaad, G.R. Investigation of physico-chemical properties for the 1-butyl-3-methylimidazolium tetrafluoroborate ([Bmim][BF_4])–diethylenetriamine (DETA) system for CO_2 capture. *J. Solut. Chem.* **2019**, *48*, 578–610. [CrossRef]
54. Harifi-Mood, A.R.; Mohammadpour, F.; Boczkaj, G. Solvent dependency of carbon dioxide Henry's constant in aqueous solutions of choline chloride-ethylene glycol based deep eutectic solvent. *J. Mol. Liq.* **2020**, *319*, 114173. [CrossRef]
55. Lin, C.-M.; Leron, R.B.; Caparanga, A.R.; Li, M.-H. Henry's constant of carbon dioxide-aqueous deep eutectic solvent (choline chloride/ethylene glycol, choline chloride/glycerol, choline chloride/malonic acid) systems. *J. Chem. Thermodyn.* **2014**, *68*, 216–220. [CrossRef]
56. Zhang, N.; Huang, Z.H.; Zhang, H.M.; Ma, J.W.; Jiang, B.; Zhang, L.H. Highly efficient and reversible CO_2 capture by task-specific deep eutectic solvents. *Ind. Eng. Chem. Res.* **2019**, *58*, 13321–13329. [CrossRef]
57. Ren, H.W.; Lian, S.H.; Wang, X.; Zhang, Y.; Duan, E.H. Exploiting the hydrophilic role of natural deep eutectic solvents for greening CO_2 capture. *J. Clean. Prod.* **2018**, *193*, 802–810. [CrossRef]
58. Siani, G.; Tiecco, M.; Di Profio, P.; Guernelli, S.; Fontana, A.; Ciulla, M.; Canale, V. Physical absorption of CO_2 in betaine/carboxylic acid-based natural deep eutectic solvents. *J. Mol. Liq.* **2020**, *315*, 113708. [CrossRef]
59. Gómez-Coma, L.; Garea, A.; Labien, Á. Hybrid solvent ([emim][Ac]+water) to improve the CO_2 capture efficiency in a PVDF hollow fiber contactor. *ACS Sustain. Chem. Eng.* **2017**, *5*, 734–743. [CrossRef]

60. Huang, Z.L.; Deng, Z.Y.; Ma, J.Y.; Qin, Y.H.; Zhang, Y.; Luo, Y.B.; Wu, Z.K. Comparison of mass transfer coefficients and desorption rates of CO_2 absorption into aqueous MEA + ionic liquids solution. *Chem. Eng. Res. Des.* **2017**, *117*, 66–72. [CrossRef]
61. Ma, C.Y.; Sarmad, S.; Mikkola, J.-P.; Ji, X.Y. Development of low-cost deep eutectic solvents for CO_2 capture. *Energy Procedia* **2017**, *142*, 3320–3325. [CrossRef]

Publisher's Note: MDPI stays neutral with regard to jurisdictional claims in published maps and institutional affiliations.

© 2020 by the authors. Licensee MDPI, Basel, Switzerland. This article is an open access article distributed under the terms and conditions of the Creative Commons Attribution (CC BY) license (http://creativecommons.org/licenses/by/4.0/).